3rd.

ube

663

AF768199

Highly Hazardous Materials Spills and Emergency Planning

HAZARDOUS AND TOXIC SUBSTANCES
A Series of Reference Books and Textbooks

Editor: Seymour S. Block
University of Florida
Department of Chemical Engineering
Gainesville, Florida

1. Highly Hazardous Materials Spills and Emergency Planning, J. E. Zajic and W. A. Himmelman

Other volumes in preparation

Highly Hazardous Materials Spills and Emergency Planning

J. E. ZAJIC
Department of Chemical and Biochemical Engineering
University of Western Ontario
London, Ontario, Canada

W. A. HIMMELMAN
Lambton College of Applied Arts and Technology
Sarnia, Ontario, Canada

MARCEL DEKKER, INC. New York and Basel

Library of Congress Cataloging in Publication Data

Zajic, James E., [Date]
Highly hazardous materials spills and emergency planning.

(Hazardous and toxic substances ; v. 1)
Includes index.
1. Hazardous substances--Accidents.
2. Environmental protection--Planning.
I. Himmelman, W. A., [Date] joint author.
II. Title. III. Series.
T55.3.H3Z33 614.8'3 78-6741
ISBN 0-8247-6622-9

Marcel Dekker, Inc.
270 Madison Avenue, New York, New York 10016

Current printing (last digit):
10 9 8 7 6 5 4 3

Printed in the United States of America

PREFACE

Toxic wastes present very special problems to all those concerned with environmental problems. In small concentrations they can usually be treated but generally they still must be segregated and handled separately. As the concentration and volume increase, the problem of storage and disposal increases. If the concentration is high enough, it may be economical to recover the chemicals involved in pure form. Highly toxic and hazardous wastes fall into a more special category. Collection, transport, storage treatment, and disposal must be considered. For some chemicals and materials, methodology and technology have not developed to the point where society can be given the assurance and the guaranty of their safety in handling. In addition, the laws and regulations relating to them are either weak or nonexistent.

Recognizing these factors, the authors have used their experience of environmental problems to prepare this book, which reviews spills of highly toxic wastes and how the problems created were resolved. Also provided is a method of evaluating a community, town, or city for potential hazards and the procedures to follow should an accident occur.

The reader should be alerted that this field is changing rapidly and that, as state laws change and as science and technology advance, the chance of a devastating spill arising from the handling and processing of highly toxic wastes will be reduced. At present, the specialized chemical properties associated with each toxicant introduces many technical problems, and these form the basis for this book.

CONTENTS

TABLES

PART 1

FIGURES

PART 1

Highly Hazardous Materials Spills and Emergency Planning

PART 1

Chapter 1

INTRODUCTION

Millions of kilograms of hazardous substances are transported daily along the highways, railways, and waterways of Canada and the United States. Some of these substances, such as cyclic pesticides, cyanides, organometallic compounds, and heavy metals are highly toxic to man and the environment, and the accidental release of even small volumes of such materials to the environment can have serious consequences. One case discussed here involves the complete destruction of the aquatic life in a lake by less than 1 kg of endrin, a pesticide.

Accidental spills and releases of hazardous substances can result from a variety of causes including highway, rail, water, industrial plant upsets, the failure of retaining dikes, storm damage, malicious acts, and even air accidents. The cost of returning the environment to an acceptable state following a spill may involve the expenditure of millions of dollars, and even after toxicants have been removed from a spill area, undesirable effects may linger in the environment for many months or years.

Increasing emphasis has been placed on environmental protection since the early 1970s. Considerable research effort has been spent on hazardous substances and their effect on the environment. Systems of preparedness have been developed and advances made on the treatment of spills. Old legislation has been amended and new acts passed. This monograph describes the progress which has been made in these areas with the emphasis on the degree of preparedness existing in the United States and Canada.

A community rating and evaluation system has been developed to define population and environmental hazards in a community so that an index or an indicator of a toxic spill crisis might be calculated.

Chapter 2

THE CLASSIFICATION OF HIGHLY HAZARDOUS MATERIALS

A clear definition of hazardous pollutants is provided by the United States Water Quality Improvement Act of 1970. Hazardous pollutants are defined as "such elements and compounds which, when discharged in any quantity or upon the navigable waters of the United States or adjoining shorelines or the waters of the contiguous zone, present an imminent and substantial danger to the public health or welfare including but not limited to fish, shellfish, wildlife, shorelines, and beaches." Under the provisions of this act, hazardous substances are classified as follows:

1. Elemental and combined forms of antimony, arsenic, beryllium, boron, cadmium, copper, chromium, lead, mercury, nickel, selenium, silver, thallium, and other elements with similar properties
2. Toxic anions such as arsenates, arsenites, chromates, cyanide fluoroaluminates, fluorides, fluorosilicates, phosphides, and others having similar properties
3. Extremely dangerous poisons (Class A) such as cyanogen and phosgene; less dangerous poisons (Class B) such as acetone, cyanohydrin, and sodium arsenite; tear gases and irritating substances (Class C) such as brombenzyl cyanide and chloracetophenone; radioactive materials (Class D) such as uranium-233 and iodine-129

4. Economic poisons including fungicides and pesticides such as DDT, aldrin, chlordane, endrin, and toxaphene
5. Any other substances that exhibit physical, chemical, biological, radioactive, or flammable properties

The Environmental Protection Act of Ontario (1971) defines a pollutant as "any contaminant or combination of contaminants present in the natural environment, or any part thereof, in excess of the maximum permissible amount, concentration, or level prescribed by the regulations." This statement defines a complete spectrum of pollutants without enlarging on hazardous substances, which may be a weakness of the act. All Canadian legislation reviewed here adheres to this broad definition.

The U.S. Federal Clean Air and Clean Water Act defines a toxic pollutant as "those pollutants or combinations of pollutants, including disease-causing agents, which, after discharge and upon exposure, ingestion, inhalation, or assimilation into any organism, either directly from the environment or indirectly by ingestion through food chains, will, on the basis of information available to the administration, cause death, disease, behavioral abnormalities, cancer, genetic mutations, physiological malfunctions (including malfunctions in reproduction), or physical deformations in such organisms or their offspring."

The 1972 amendments to the U.S. Federal Water Pollution Control Act (P.L. 92-500, 85 Stat. 816-1972) defines toxic pollutants as those that "cause death, disease, behavioral abnormalities, cancer, genetic mutations, physiological malfunctions (including malfunctions in reproduction) or physical deformations in such organisms or their offspring."

Many classification systems have been developed for hazardous substances considering the rather broad meaning of the word "hazard." Included are flammable liquids and solids, combustible liquids, oxidizing materials, explosives, etiological materials, compressed gases, cryogenic materials, radioactive materials, poisons, corrosive materials, and molten substances. It would be too formidable a task to cover all aspects of each form of hazard a toxic compound exhibits in the environment; this monograph is therefore directed mainly toward highly hazardous substances as they relate to the aquatic environment. It is believed that a society that can effectively cope with a voluminous spill of highly hazardous material can cope with spills of less toxic materials.

Hazardous materials are classified into the following groups by the Code of U.S. Federal Regulations, Title 49, Transportation, Parts 100-199.

Explosives (Classes A, B, and C)
Radioactive materials
Poisons (Classes A, B, and C)
Corrosive liquids

Details pertaining to the latter two classifications are as follows:

Class A Poisons:

"Extremely dangerous poisons. Poisonous gases or liquids of such a nature that a very small amount of the gas or vapor of the liquid mixed with air is dangerous to life." (S 173.326)

Class B Poisons:

"Less dangerous poisons. Substances, liquids or solids (including pastes and semisolids), which are known to be so toxic to man as to afford hazard to health during transportation or which, in the absence of adequate data on human toxicity, are presumed toxic to man." (S 173.343)

Class C Poisons:

"Tear gas or irritating substances. Liquid or solid substances which, upon contact with fire or when exposed to air, give off dangerous or intensely irritating fumes, but not including any poisonous article, Class A." (S 173.381)

Corrosive Liquids:

"Corrosive liquids that by contact will cause severe damage to living tissue, will damage or destroy other freight by chemical action or will be liable to cause fire when in contact with inorganic matter or certain chemicals." (S 173.240)

A report to the United States Coast Guard, prepared by the Committee on Hazardous Materials of the National Research Council [1], suggested a classification of hazardous materials based on fire and health hazards, water pollution potential, and reactivity. This proposed classification was arranged in five categories from Grade 0, which defined substances having a very low environmental hazard rating, to Grade 4, substances that are highly hazardous. Grades 3 and 4 materials are the type of substances that fit the requirements of "highly hazardous" and these classifications are defined in Table 2-1. The values for LD_{50} are based on ratings in the Handbook of Toxicology [2] and TL (threshold limit) values are from the industrial hygienist publication Threshold Limit Values [3].

A list of highly toxic substances was proposed by the U.S. Environmental Protection Agency (EPA), Federal Register, December 1973 [4]. Some of these compounds are listed below and are considered to be highly toxic to the water environment and, if discharged into a body of water, will result in an extreme environmental upset of long duration plus a costly cleanup program.

Aldrin-dieldrin, endrin, toxaphene
Benzidine
Cadmium and cadmium compounds
Cyanides
DDT, DDE, DDD
Mercury and mercury compounds
Polychlorinated biphenyls

The toxicity of the following are under assessment:

Compounds of arsenic, selenium, chromium, lead, zinc, beryllium, nickel, and antimony
Asbestos

Table 2-1

Highly Hazardous Materials, Biological Classification

	Classification	
	Grade 3	Grade 4
LD_{50}[a], mg/kg	50 to 500	<30
Threshold limit value (TLV), ppm	1 to 100	<1
Toxicity	Moderate	Severe
Irritation	Liquids produce second degree burns after a few minutes exposure. Vapors cannot be tolerated.	Liquids produce second and third degree burns. Vapors can produce permanent injury or death.
Reactivity	Hazardous substances may be produced with water and vigorous self-reactions occur requiring stabilizing.	Hazardous substances produced with water. Self-reactions may produce detonations.

Source: U.S. House Document No. 92-70, 1971, Control of Hazardous Polluting Substances, Section 11-C.

[a]1-hr exposure.

Chlordane, lindane, heptachlor, parathion and methyl parathion

Hydroquinone, camphor, o-chlorophenol, acridine, and α-naphthol and di-n-butyl phthalate

Water-soluble materials are difficult to retain in the aqueous environment and represent a challenge to emergency planning. The Battelle Memorial Institute, in a study for the EPA [5], defined a priority ranking system for primarily water-soluble substances based on their properties, the annual quantities shipped, and the probabilities for spillage, depending on the mode and route of transportation. Figure 2-1 assesses both soluble and insoluble substances.

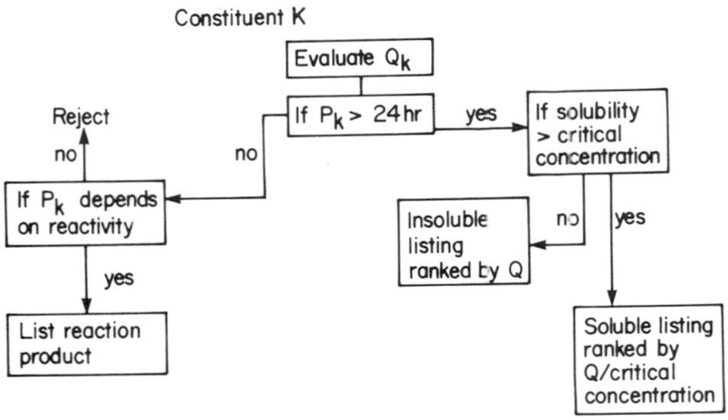

Figure 2-1, Hazardous materials priority ranking system. K = a compound; Q_k = quantity of compound potentially spilled into water; P_k = persistence time of spill. (Compounds with low boiling points, a specific gravity value less than 1, and low solubility will have a P_k value < 24 hr.) Source: G. W. Dawson, A. J. Shuckrow, and W. H. Swift, Control of Spillage of Hazardous Polluting Substances, for FWQA, 1970, Document 15090 FOZ 10/70, p. 18.

Concern here is directed to the soluble substances which must be assessed by the ratio of Q to the critical concentration. Critical concentration is defined as the lowest threshold concentration of four water quality parameters, namely human, aquatic, and plant toxicity, and aesthetic effects. Values for certain highly hazardous materials are summarized in Table 2-6.

The value of Q is given by:

$$Q = Q_1[(P_w f_w) + (P_r f_r a_r) + (P_t f_t a_t)]$$

where Q_1 = annual production quantity of the compound

P_w = probability of any barge shipment resulting in an accident = 0.0028 (for the United States; number derived from the total number of barge accidents related to hazardous materials and the total number of barge chemical shipments in 1968)

f_w = percentage of compound shipped by water

P_r = probability of any rail shipment resulting in an accident:

= 0.0011 (number derived from the total number of rail accidents related to hazardous materials and the total number of rail chemical shipments in 1968)

f_r = percentage of compound shipped by rail

a_r = fraction of major railways near surface waters

= 0.3 (number derived by mapping rail routes and waterways between major cities in each region of the United States)

P_t = probability of any truck shipment resulting in an accident

= 0.019 (number derived from the total number of truck accidents related to hazardous materials and the total number of truck chemical shipments in 1968)

f_t = percentage of compound shipped by truck

a_t = fraction of major highways near surface waters

= 0.3 (number derived by mapping truck routes and waterways between major cities in each region of the United States)

Of 257 materials rated by the "Battelle" priority ranking system, the top 15 hazardous substances in the United States are as follows:

Ranking	Substance
1	phenol
2	methyl alcohol
3	cyclic rodenticides
4	acrylonitrile
5	chlorosulfonic acid
6	benzene
7	ammonia
8	cyclic insecticides
9	phosphorous pentasulfide
10	styrene
11	acetone cyanohydrin
12	chlorine
13	nonyl phenol
14	DDT
15	isoprene

Nearly all of these materials are transported in large quantities also in Canada. It is expected that Battelle's priority system closely approximates the situation in Canada.

Environment Canada [6] has classified hazardous polluting substances in the area of the lower Great Lakes basin using the relationship:

$$R_c = Q_s + 0.024\ Q_r + 0.017\ Q_t + 0.10\ Q_p + 0.054\ Q_{st}$$

where R_c = the combined risk potential

Q = the category rating for the quantity of material shipped or stored

s, r, t, p, and st = ship, rail, truck, pipeline, and storage, respectively

Risk potential values for selected hazardous materials are listed in Table 2-2. The risk potential value has been incorporated with an <u>ecological effect rating</u> (E) and a <u>use and distribution rating</u> (U) in the following relationship where OR is the <u>overall rating value</u>.

$$OR = \sqrt{EUR_c}$$

The ecological effect rating is given by:

$$E = TP = C^n(B + b)$$

where $T = C^n$ = the ecological index

$P = B + b$ = the persistence index

n = solubility factor based on the ratio of the 4-day LD_{50} of the material to the solubility of the material and has assigned values ranging from 0.5 to 1.0

C = basic toxicity rating and is defined as follows:

4-day LD_{50} Concentration	C Value
less than 0.1 mg/liter	4
0.1 to 100 mg/liter	$3 - \log_{10} LD_{50}$[a]
greater than 100 mg/liter	1

B = persistence rating, based on the probable half-life of the substance having assigned values of 1.0, 1.5, and 2.0

Table 2-2

Overall Risk Estimations for Selected Hazardous Substances

Substance	R_c[a]	U[a]	E[a]	OR[a] = (EUR_c)
Metallic salts, inorganic acids	5.82	4[b]	10[b]	14.5
Plastic materials	5.02	5[b]	4[b]	10.0
Rubber products	5.01	5[b]	2[b]	7.1
Hydrocarbons and derivatives	5.00			
Phenols, ethers, aldehydes, ketones	4.99			
Organic chemicals	4.98			
Sulfuric acid	4.84	5.4	4[b]	10
Inorganic bases	4.82			
Sodium hydroxide	4.80	5.0	4[b]	9.8
Chemical specialties	4.69			
Motor gasoline	7.36	5.4	4[b]	12.6
Diesel fuel oil	8.03	5.8	2[b]	9.6
Crude oil	6.20	4.1	5[b]	11.2
Hydrochloric acid	4.81	3.6	4[b]	8.3
Fertilizers	5.53			

Source: D. M. Gorber and J. R. Monteith, A Rating of Various Chemicals as Specific Hazards in the Great Lakes, Proc. 1974 National Conf. Control of Hazardous Material Spills, San Francisco, August 25-28, 1974, pp. 25-30.

[a]R_c = combined risk potential; U = use and distribution rating; E = ecological effect rating; OR = overall rating value.

[b]Estimated by authors.

b = a physical state correction with assigned values of 1.0, 0.5, 0, -0.5, and -1.0

Ecological effect ratings are summarized as follows:

Rating value	Conditions
0 to 1.5	Minimal hazards, low toxicity and low persistence

1.5 to 4.0	Slight hazards, combination of low toxicity and high persistence
4.0 to 6.5	Moderate hazards, moderate toxicity, and moderate to high persistence
6.5 to 9.0	Highly hazardous materials, moderate to high toxicity, and high persistence
9.0 to 12.0	Extreme hazards, high toxicity, and very high persistence

Based on information given by Environment Canada, the values have estimated the overall rating for certain of the materials listed in Table 2-2.

The highly hazardous substances of prime concern are those listed in Table 2-3. The materials listed in this table fall into one or more of the following categories:

1. Highly toxic substances that represent an extreme hazard to man and the environment
2. Highly corrosive substances
3. Highly hazardous water soluble substances
4. Very hazardous substances that represent high volumes with regard to manufacturing and handling facilities and transportation

Radioactive substances are discussed in Chapter 7, including radioactive pollutants and their sources, the effects of radioisotopes on living organisms, nuclear accidents, and the care that government and industry have taken to protect the environment.

Not too much emphasis is placed on flammable substances and explosives. Pathological wastes are not included here because of their presumably low volume and the fact that they are generally incinerated at their source. All hospital wastes should be treated and all pathogens killed before they are discharged or released from the hospital area. Thus, all toxicants must be examined in relationship to how they affect the environment, their solubility, and whether they can be contained.

Table 2-3

Highly Hazardous Materials, Group Classification of Materials Emphasized in This Study

1. Highly Toxic Substances
    - Economic poisons (fungicides, herbicides, and pesticides)
    - Polychlorinated biphenyls
    - Cyanides
    - Metallic compounds
2. Toxic Gases
    - Phosgene
    - Hydrogen sulfide
    - Chlorine
3. Highly Corrosive Liquids
    - Strong acids (sulfuric, hydrochloric, phosphoric, etc.)
4. Others
    - Chemicals (phenol, methyl alcohol, acrylonitrile, benzene, ethylene oxide, ethylenimine, etc.)
    - Industrial wastes

Classification and physical properties of hazardous and highly hazardous materials are shown in Tables 2-4 through 2-7. Some definitions are provided in Table 2-8. Certain correlations are evident, such as health hazard ratings and the threshold limit values of materials, and vapor pressure and flash point data versus flammability ratings. The data on solubility in water indicate the damages that might occur to the aquatic environment if such materials spill into creeks or rivers.

Toxic ratings, detection limits, and the solubility of the more common highly hazardous materials are listed in Table 2-7. Information is presented on fungicides, herbicides, and pesticides as well as cyanides, metallic compounds, and selected chemicals. Many of the compounds have very low TLm values (high toxicity) for marine (page 34)

Table 2-4

Classification of Hazardous and Highly Hazardous Materials, Hazards to Humans, and General Information

---

Definition of Ratings and Terms

Nature of Associated Hazards

1. Flammable substance
2. Oxidizing substance, reacts with reducing agents
3. Emits a toxic gas or vapor
4. Emits an irritating gas or vapor
5. Emits a narcotic gas or vapor
6. Gas or vapor is not dangerous other than displacing air
7. Causes skin irritations or burns
8. Toxic substance
9. Explosive material under certain conditions

Hazard Ratings

Health

0 None

1 Minor

2 Moderate, could cause temporary incapacitation or injury

3 Severe, short exposure may cause serious injury

4 Extreme, short exposure may cause death

Flammability

0 None, material does not burn

1 Minor, material must be preheated to ignite

2 Moderate, moderate heating is required for ignition and volatile vapors are released

3 Severe, material ignites at normal temperature

4 Extreme, very flammable substance that readily forms explosive mixtures

Reactivity

0 None, stable when exposed to fire

1 Minor, unstable at high temperatures or pressures and may react with water

Table 2-4 (continued)

Reactivity

2	Moderate, unstable but does not explode; may form explosive mixtures with water
3	Severe, explodes if heated or water added
4	Extreme, readily explosive under normal conditions

Threshold Limit

The concentration in air to which people can be exposed for a period of 8 hr without injury. The values for gases and vapors are in parts of gas or vapor per million parts of air by volume at 25° C and one atmosphere at a pressure of 101.3 kPa (1 atm). The values for solids are in mg of substance per m^3 of air.

General Description

wh	white
cryst	crystals
n col	colorless
od	odor
powd	powder

Flash Point

The lowest temperature at which material gives off vapors that can be ignited in the presence of a spark.

Autoignition Temperature

The temperature at which the substance ignites spontaneously.

Vapor Pressure

Expressed in Torr (mm of mercury) at the temperature indicated (Torr x 133.29 = Pascal).

Vapor Density

A value greater than 1 indicates that the gas or vapor is heavier than air. Air is equivalent to unity.

Solubility in Water

sol	Soluble, 5-50 g per 100 ml
v sol	Very soluble, greater than 50 g per 100 ml
s sol	Slightly soluble, less than 5 g per 100 ml
insol	Insoluble
m	Miscible in all proportions
d	Decomposes in water

Table 2-5

Classification and Physical Properties of Hazardous and Highly Hazardous Materials, Solids (See Table 2-4 for hazard rating definition and units.)

Classification

Substance	Synonym	Nature of Associated Hazards	Hazard Ratings: Health	Hazard Ratings: Flammability	Hazard Ratings: Reactivity	Threshold Limit
Aluminum chloride		3,7	3	0	2	
Ammonium nitrate		1,2,3,9	2	1	3	
Antimony trichloride;	antimonious chloride	3,4,7,8				
Arsenic compounds		8	4			0.5
Benzoic acid		1				
Calcium carbide:	Carbide	1,3,7	1	4	2	
Calcium oxide:	Quicklime	7	1	0	1	5
Calcium hypochlorite:	Bleaching compound	2,3,7	2	1	2	
Camphor:	2-Camphanone	4,5,6,7	2	2	0	2
Chromic acid:	Chromic anhydride	4,5	1	0	1	0.1
Copper compounds		5,6	1	0	1	
1,2-Dinitrobenzene		1,4,7	3	1	4	1.0
Ferric chloride:	Molysite	5				
Lead compounds		4,7	1	0	1	
Mercury compounds		5,6				

Monochloroacetic acid	Chloroethanoic acid	5				
Naphthalene		7	2	2	0	50
Nitrocellulose:	Cellulose nitrate	1,4,7	1	3	3	
Oxalic acid:	Ethanedioic acid	6				1.0
Phenol:	Carbolic acid	1,5,7	3	2	0	19
Potassium hydroxide:	Potassium hydrate	2,3,5,6	3	0	1	
Sodium		1,2,4,5	3	1	2	
Sodium chlorate:	Soda chlorate	2,8	1	0	2	
Sodium fluoride:	Villaumite	5,6				
Sodium hydroxide:	Caustic soda	2,3,5,6	3	0	1	2.0
Sodium peroxide		2,3,4,5,6,7	3	0	2	
Sulfur		4,7	2	1	0	
Trinitrotoluene:	TNT	1,4,7	2	4	4	1.5
Zinc chloride		5	2	0	2	1.0

Physical Properties

Substance	General Description	Flash Point, °C	Auto-ignition, °C	Density, g/ml	Melting Point, °C	Solubility in Water
Aluminum chloride	Yellowish-white cryst			2.44	192.4	Sol
Ammonium nitrate	N col, cryst			1.72	169.6	V sol

Table 2-5 (continued)

Substance	General Description	Flash Point, °C	Auto-ignition, °C	Density, g/ml	Melting Point, °C	Solubility in Water
Antimony trichloride	N col, transp cryst			3.14	73.4	V sol
Arsenic compounds						
Benzoic acid	White powd	120	574	1.32	121.7	S sol
Calcium carbide	Gray cryst			2.22	2300	d
Calcium oxide	N col, cryst			3.37	2580	V sol
Calcium hypochlorite	White powd			2.35	100 (d)	Sol
Camphor	White cryst mass	66	466	0.99	174	S sol
Chromic acid	Purple-red cryst			2.70	196	Insol
Copper compounds						
1,2-Dinitrobenzene	N col to yellow needles	150		1.57	118	Insol
Ferric chloride	Brownish-black solid			2.80	282	Sol
Lead compounds						
Mercury compounds						
Monochloroacetic acid	N col cryst			1.58	63 (d)	V sol

Naphthalene	White flakes	80	525	1.14	80.1	Insol
Nitrocellulose	White powd	27		1.66	160	Insol
Oxalic acid	Transp, n col cryst			1.65	101	Sol
Phenol	White cryst mass	79.4	715	1.07	40.6	Sol
Phosphorus (white)	Col to yellow, waxy	Spontaneous ign.		1.82	44.1	S sol
Picric acid	Yellow cryst	150	300 expl	1.76	121.8	S sol
Potassium hydroxide	White pellets, flakes			2.04	360	V sol
Sodium	Silvery white metal			0.97	97.8	D
Sodium chlorate	N col cryst			2.49	248	V sol
Sodium cyanide	White cryst or powd				563	Sol
Sodium fluoride	Clear lustrous cryst			2.56	992	V sol
Sodium hydroxide	White pellets, flakes			2.12	318.4	V sol
Sodium peroxide	White powd			2.81	460 (d)	Sol
Sulfur	Rhombic yellow cryst	207	232	2.07	112.8	Insol
Trinitrotoluene	N col cryst	Explodes		1.65	80.7	Insol
Zinc chloride	White cryst			2.91	262	Sol

Table 2-6

Classification and Physical Properties of Hazardous and Highly Hazardous Materials, Liquids and Gases

Substance	Synonyms	Nature of Associated Hazards	Hazard Ratings			Threshold Limit
			Health	Flammability	Reactivity	
Classification: Liquids						
Acetaldehyde	Acetic anhydride	1,4,7,9	2	4	2	200
Acetic acid		1,4,7	2	2	1	10
Acetone	Dimethyl ketone	1,5,9	1	3	0	1000
Allyl alcohol	Vinyl carbinol	1,4,9	3	3	1	2
Amyl alcohol	Pentyl alcohol	1,4,9	1	3	0	125
Aniline		1,3,7,8	3	2	0	5
Benzaldehyde	Benzoic aldehyde	1	2	2	0	
Benzene	Benzol	1,3,8,9	2	3	0	<25
Bromine		2,3,7	4	0	1	0.1
Butyl alcohol	Butanol	1,5,7,9	1	3	0	100
Butyric acid	Butanoic acid	1	2	2	0	
Carbon disulfide		1,3,8,9	2	3	0	20
Carbon tetrachloride	Tetrachloromethane	3,8				10
Chlorobenzene	Phenyl chloride	1,5,9	2	3	0	75
Cresol	Cresylic acid	1,3,7,8	2	2	0	5

Cyclohexane		1,5,7,9	1	3	0	300
Fuel oil		1,9	0	2	0	
Diethylamine		1,7	3	3	0	25
Ethyl acetate		1,4,9	1	3	0	400
Ethyl alcohol	Ethanol	1,4,9	0	3	0	1000
Ethylene glycol	1,2-Ethanediol	1	1	1	0	
Ethyl ether		1,5,9	2	4	1	400
Formalin	Formaldehyde	1,4,7	2	2	0	5
Formic acid	Methanoic acid	1,3,7,8	3	2	0	5
Furfural	2-Furaldehyde	1,4,8,9	1	2	1	5
Gasoline		1,6,9	1	3	0	
Hydrazine		1,3,7,9	3	3	2	1
Hydrochloric acid (conc.)	Muriatic acid	3,7,8	3	0	0	<5
Hydrocyanic acid	Hydrogen cyanide	1,3,7,8,9	4	4	2	10
Hydrofluoric acid		3,4,7,8	4	0	0	3
Hydrogen peroxide (<50%)		2,7,9	2	0	3	1
Isopropyl alcohol	Isopropanol	1,4,5,9	1	3	0	400
Kerosene		1,9	0	2	0	
Mercury	Quicksilver	3,8				0.1
Methyl alcohol	Methanol	1,5,8,9	1	3	0	200

Table 2-6 (continued)

Substance	Synonyms	Nature of Associated Hazards	Hazard Ratings			Threshold Limit
			Health	Flammability	Reactivity	
Ketone (methyl ethyl)	MEK, 2-butanone	1,4,9	1	3	0	200
Naphtha		1,9	0	2	0	100
Nitric acid (conc.)	Aqua fortis	2,3,4,7,8	2	0	1	2
Nitrobenzene	Oil of mirbane	1,3,8,9	3	2	0	1
Nitroglycerin	Glycerol trinitrate	1,8,9	2	2	4	0.2
Nitromethane		1,3,9	1	3	4	35
Perchloric acid		2,4,9	3	0	3	
Petroleum ether	Benzine	1,5,9	1	4	0	500
Phosphoric acid		3,7,9				1
Phosphorus oxychloride		3,7,8				
Phosphorus trichloride	Phosphorous chloride	3,7,8	3	0	2	0.5
n-Propyl alcohol	Propanol-1	1,5,9	1	3	0	200
Pyridine		1,3,7,8,9	2	3	0	5
Styrene	Phenyl ethylene	1,5,9	2	3	2	< 100
Sulfuric acid (conc.)	Oil of vitriol	2,3,7	3	0	1	1
Thionyl chloride	Sulfurous oxychloride	3,7,8				

Toluene	Methylbenzene, toluol	1,4,9	2	3	0	200
Xylene	Dimethylbenzene, xylol	1,8,9	2	3	0	100
Classification: Gases						
Acetylene	Ethyne	1,5,9	1	4	3	
Ammonia		1,4,7	3	1	0	50
Arsine	Arsenic hydride	1,3,7,8				0.05
Boron trifluoride	Boron fluoride	3	3	0	1	< 1
Butane	n-Butane	1,9	1	4	0	
Carbon monoxide		1,3,9	2	4	0	50
Chlorine		1,2,4,7,9	3	0	1	< 1
Ethane		1,6,9	1	4	0	
Fluorine		1,2,4,7	4	0	3	0.1
Hydrogen		1,6,9	0	4		
Hydrogen cyanide	Hydrocyanic acid	1,3,8,9	4	4	2	10
Hydrogen sulfide		1,4,5,9	3	4	0	10
Nitrogen dioxide	Nitrogen tetroxide	2,3,4,7,8	3	0	1	< 5
Oxygen		2,9				
Ozone		2,4,9				0.1
Phosgene		3				0.1
Phosphine	Hydrogen phosphide	1,3,9				0.1
Propane		1,5,9	1	4	0	1000
Sulfur dioxide		4,7	3	0	0	5

Table 2-6 (continued)

Substance	General Description	Flash Point, °C	Auto-ignition, °C	Vapor Pressure, Torr/°C	Boiling Point, °C	Vapor Density	Solubility in Water
Physical Properties: Liquids							
Acetaldehyde	N col, pungent od	-37.8	185	740/20	20.8	1.52	M
Acetic acid	N col, pungent od	42.8	426	11.4/20	118.1	2.07	M
Acetone	N col, mintuke od	-17.8	538	400/39.5	56.5	2.00	M
Allyl alcohol	Pungent od	21.1	377	23.8/25	96	2.00	Sol
Amyl alcohol	N col	-11.7	572	1/13.6	137.8	3.04	Insol
Aniline	N col, oily	70	769	15/77	184.4	3.22	Sol
Benzaldehyde	N col	64.4	191	1/26.2	179	3.65	S sol
Benzene	N col	-11.1	562	100/26.1	80.1	2.77	S sol
Bromine	Dark red			77.3/4	58.7	5.5	S sol
Butyl alcohol	N col	28.9	365	5.5/20	117.5	2.55	Sol
Butyric acid	Rancid od	71.7	452	0.43/20	163.5	3.04	Sol
Carbon disulfide	N col, unpleasant od	-30	100	400/28	46.5	2.64	Insol
Carbon tetrachloride	N col			100/23	76.8		S sol
Chlorobenzene	N col	29.4	638	10/22.2	131.7	3.88	Insol
Cresol	N col	81	599	1/38.5	191	3.72	Sol
Cyclohexane	N col, pungent od	-20	260	100/60.8	80.7	2.90	Insol
Fuel oil		37.8	257				Insol

Diethylamine	N col	< 0	312		134	2.5	V sol
Ethyl acetate	N col, fragrant od	4.4	427	100/27	77.2	3.04	Sol
Ethyl alcohol	N col, fragrant od	12.8	422	40/19	78.3	1.59	M
Ethylene glycol	N col	111.1	412	0.06/20	197.5	2.14	M
Ethyl ether	N col	-45	180	442/20	34.6	2.56	Sol
Formalin	Pungent od	85	403		101		Sol
Formic acid	N col, pungent od	68.9	601	43/25	100.8		M
Furfural	N col to yellow	60	316	1/19	161.7	3.31	Sol
Gasoline		< -40	257			3.0	Insol
Hydrazine	N col, fuming	52.2	270	14.4/25	113.5	1.1	V sol
Hydrochloric acid (conc.)	N col, fuming						M
Hydrocyanic acid	N col, faint almond od	-17.8	538	400/9.8	25.7	0.932	M
Hydrofluoric acid	N col, fuming				19.4		M
Hydrogen peroxide (< 50%)	N col				107		M
Isopropyl alcohol	N col	11.7	399	44/25	82.3	2.97	M
Kerosene	N col to pale yellow	~37	~229		175 (min)	4.5	Insol
Mercury	Silvery, heavy liquid			1/38.4	356.9		Insol
Methyl	N col	11.1	446	160/30	64.8	1.11	M
Methyl ethyl ketone	N col	-6.1	474	71.2/20	79.6	2.41	V sol
	N col to dark	~29	~232		~149-216		Insol
Naphtha	Suffocating od,						
Nitric acid (conc.)	N col to yellow				86		M

Table 2-6 (continued)

Substance	General Description	Flash Point, °C	Auto-ignition, °C	Vapor Pressure, Torr/°C	Boiling Point, °C	Vapor Density, g/ml	Solubility in Water
Nitrobenzene	Yellow, oily	87.8	482	1/44.4	210.9	4.25	Insol
Nitroglycerin	N col to yellow	Explodes		1/121	Explodes	7.84	Insol
Nitromethane	Oily	44.4	379	27.8/20	108	2.11	Sol
Perchloric acid	N col, fuming				200		M
Petroleum ether	N col, volatile	~ -18			~40-80	2.5	Insol
Phosphoric acid	N col, syrupy			0.03/20			V sol
Phosphorus oxychloride	Yellow			40/27.3	105.1	5.3	D
Phosphorus trichloride	N col, fuming			100/21	74.2	4.75	D
n-Propyl alcohol	N col	15	371	20.8/25	97.2	2.07	V sol
Pyridine	N col, sharp od	20	482	20/25	115.3	2.73	M
Styrene	N col	32	490		~145.5	3.6	Insol
Sulfuric acid (conc.)	N col, oily			1/146	330		M
Thionyl chloride	N col to red			100/21	78.8		D
Toluene	N col	4.4	506	30/20	110.4	3.14	Insol
Xylene	N col	28.9	528	10/28.3	139	3.66	Insol

Physical Properties: Gases

Acetylene	Garlic od, N col	-17.8	299		-84.0	0.91	S sol
Ammonia	Pungent od, N col		646	10 atm/25.7	-33.4	0.6	V sol
Arsine	Garlic od, N col				-55	2.66	S sol
Boron trifluoride	N col			27.9/30	-110.7		Sol
Butane	N col	-60	404	1823/25	0.5	2.05	V sol
Carbon monoxide	N col, odorless		608	760/19	-191.3	1.0	Sol
Chlorine	Greenish-yellow			3.66/0	-34.5	2.49	S sol
Ethane	N col, odorless		514		-88.6	1.10	Insol
Fluorine	Pale yellow				-187	1.69	D
Hydrogen	N col		584		-252.8	0.069	S sol
Hydrogen cyanide	Almond od, N col		37.7	400/9.8	25.7	0.932	M
Hydrogen sulfide	Od rotten eggs, N col		260		-60.4	1.189	V sol
Nitrogen dioxide	Reddish-brown			400/80	21		Sol
Oxygen	N col, odorless				-18.3	1.429	S sol
Ozone	N col				-111.1		Sol
Phosgene	Od hay, N col			1180/20	8.3	3.4	D
Phosphine	N col				-87.5		S sol
Propane	N col	-104	467	8.8/20	-42.1	1.56	Sol
Sulfur dioxide	N col			2538/21	-10.0	2.264	V sol

Source: National Fire Protection Association, Publication 1968-8, Fire Protection Guide on Hazardous Materials, American Conference of Governmental Industrial Hygienists.

Table 2-7

Critical Values for Highly Hazardous (Toxic) Substances

Substance	Toxicity: Rat, mg/kg body wt	Toxicity: Humans, mg/liter	Toxicity: Fish, mg/liter	Toxicity: Plants, mg/liter	Field Detection Limit, mg/liter	Water Solubility, mg/liter
Fungicides						
Captan	10,000		0.30		0.3	Insol
Dichloronaphthoquinone	1,500		0.07			0.1
Mercury compounds		0.005 (Hg)			0.5 (Hg)	
Nabam	395		0.5		1,000	
Pentachlorophenol	78		0.2		0.1	Insol
Thiram	350		0.79		0.05	Insol
ZDD			0.008		0.05 (Zn)	
Herbicides						
2,4-D	500		100	20	2	Sol
CMU	3,500	180	40	8	5	Sol
Diuron	3,400		5			42
Endothal	35		140	100	5	Sol
MCP			15		10	
Silvex (2,4,5-T)	650		1.0		0.1	S sol
TBA (2,3,6)	300		150		50	

TCA (2,4,5-T acid, esters, salts)	300		10	0.1	
Pesticides					
Aldrin-toxaphene group	40		0.023	0.5	Insol
Benzine hexachloride (lindane)	125		0.018	100	10
Chlordane	340		0.01		Insol
Chlorthion	550		0.0045	1 (PO_4^{3-})	Insol
DDD (TDE or rhothane)	3,400		0.013	0.001	Insol
DDT	250		0.0023	0.001	0.2
Diazinon	100		0.004	1 (PO_4^{3-})	
Dicapthon	460		2	1 (PO_4^{3-})	
Dieldrin	40	0.1/5g	0.003	0.5	Insol
Dipterex	400		50	1 (PO_4^{3-})	Sol
Endrin	10		0.001		Insol
Guthion	11		0.01	1 (PO_4^{3-})	33
Heptachlor	90		0.019		Insol
Isodrin	7		0.0015	0.5	
Kelthane	575		0.5		
Malathion	1,000		0.02	1 (PO_4^{3-})	45
Metasystox	40		5	1 (PO_4^{3-})	S sol
Methoxychlor	5,000		0.025		S sol
Methyl parathion	3	150	0.04	1,000	
Nicotine	55	>10	8	20	

Table 2-7 (continued)

Substance	Toxicity: Rat, mg/kg body wt	Toxicity: Humans, mg/liter	Toxicity: Fish, mg/liter	Toxicity: Plants, mg/liter	Field Detection Limit, mg/liter	Water Solubility, mg/liter
Ovotran	2,000		10		5	Insol
Phosdrin	6		0.017		1 (PO_4^{3-})	V sol
Pyrethrum	200		2			
Rotenone	132		0.022			17
Schradan	9		120	1,000	1 (PO_4^{3-})	
Sevin	500		5		5	1,000
Sulfoxide			0.75			
Systox	2.5		3.6		1 (PO_4^{3-})	Insol
TEPP (HETP)	1.2		1.0		1 (PO_4^{3-})	V sol
Thiodan	90		0.01			
Toxaphene	69		0.005		0.5	1.5
Kuron	500		1.23			Insol
Parathion	2		0.001		1,000	20
Vapam	800		1		5	Sol
Cyanides						
Acrylonitrile	90	0.01 (CN^{1-})	15		10	Sol
HCN (cyanic acid)		0.01 (CN^{1-})	0.05	< 100	0.03 (CN^{1-})	M

KCN (potassium cyanide)		0.01(CN^{1-})	0.4		0.03 (CN^{1-})	V sol
Acetone cyanohydrin		0.01(CN^{1-})	1-100		0.03 (CN^{1-})	V sol
Metallic compounds						
Mercuric salts (Hg)		0.005	0.02+	37($HgCl_2$)	0.01	Sol
Cadmium salts (Cd)			6	50	0.04	Sol
Tetraethyl lead (Pb)		0.05	0.2	50	0.05	Insol
Lead salts (Pb)		0.05	0.5-25	50	0.05	Varies
Corrosive Liquids						
Sulfuric acid			10		1	M
Hydrochloric acid			10		0.4	M
Hydrofluoric acid		1	40		0.05	M
Nitric acid		15	1			M
Others						
Phenol			0.1	1,000	0.001	Sol
Methyl alcohol	9.1	0.01	250		3.5	M
Benzene	5,600		5		0.5	820
Ethylene oxide			100-1,000			M
Ethylenimine	15		1-100		10	M

Source: G. W. Dawson, A. J. Shuckrow, and W. H. Swift, Control of Spillage of Hazardous Polluting Substances, 1970, for FWQA, Document 15090 FOZ 10/70, Washington, D.C., 1970, p. B-2.

Table 2-8

Definitions of Some Highly Toxic Substances

Aldrin-Dieldrin:	
Aldrin:	1,2,3,4,10,10-hexachloro-1,4,4a,5,8,8a-hexahydro-1,4-endo-exo-5,8-dimethanonaphthalene
Dieldrin:	1,2,3,4,10,10-hexachloro-6,7-epoxy-1,4,4a,5,6,7,8,8a-octahydro-1,4-endo-exo-5,8-dimethanonaphthalene
Cadmium:	the element cadmium and all its compounds
Cyanide:	any cyanide compound that produces free cyanide ion or molecular HCN in effluents
DDT:	compounds of DDT, DDD, and DDE
DDT:	1,1,1-trichloro-2,2-bis(p-chlorophenyl)ethane
DDD:	1,1-dichloro-2,2-bis(p-chlorophenyl)ethane
DDE:	1,1-dichloro-2,2-bis(p-chlorophenyl)ethylene
Endrin:	1,2,3,4,10,10-hexachloro-6,7-epoxy-1,4,4a,5,6,7,8a-octahydro-1,4-endo-endo-5,8-dimethanonaphthalene
Mercury:	the element mercury and all compounds of mercury
PCBs:	materials containing the biphenyl group which have been chlorinated to varying degrees; there are 210 possible different PCB compounds
Toxaphene:	chlorinated camphene; approximate formula is $C_{10}H_{10}Cl_8$

forms. The pesticide endrin has a TLm value of 0.001 mg/liter for fish, which means that a very small spill can produce major damage to the aquatic environment.

REFERENCES

1. U.S. House Document 92-70, Control of Hazardous Polluting Substances, A Report on Control of Hazardous Polluting Substances pursuant to Section 12(g) of the Federal Water Pollution Control Act as amended, U.S. Govt. Print. Off., Washington, 1971.

2. Committee on the Handbook of Biological Data of the NAS-NRC Division of Biology and Agriculture, Handbook of Toxicology, Acute Toxicities of Solids, Liquids and Gases to Laboratory Animals, Vol. I, Saunders, Philadelphia, 1956.

3. Annual Publ. by NAS-NRC Committee on Technology, Threshold Limit Values. Threshold Limit Values Committee of the American Conference of Government Industrial Hygienists, Cincinnati, Ohio.

4. Environmental Protection Agency, Water Programs, Proposed Toxic Pollutant Effluent Standards, Fed. Register, 38, No. 242, Part II, Dec. 27, 1973.

5. G. W. Dawson, A. J. Shuckrow, and W. H. Swift, Control of Spillage of Hazardous Polluting Substances, U.S. Federal Water Quality Administration, Document 15090 FOZ 10/70, Washington, 1970.

6. D. M. Gorber and J. R. Monteith, A Rating of Various Chemicals as Specific Hazards in the Great Lakes, Proc. 1974 National Conf. Control of Hazardous Material Spills, San Francisco, August 25-28, 1974, pp. 25-30. (Library of Congress Catalog No. 74-16083.)

Chapter 3

ENVIRONMENTAL EFFECTS OF SPILLS

Spills may impair the quality of water, land, or air, or all three simultaneously. A land spill of a nonvolatile, toxic substance generally has less impact on the environment than a spill of a highly volatile, toxic substance or a substance that has entered a waterway. Land spills can normally be contained and the contaminated soil removed to a disposal area, localizing the effects of the incident. Where air or water contamination is involved, the fluidity of these media favor the transport of highly hazardous substances over a wider area with possible dangerous consequences. Little more can be provided following a serious atmospheric release other than the evacuation of personnel and stopping the leak at its source. Once released to the atmosphere, vapors are rapidly dispersed by diffusion and wind currents and containment is impossible.

Simmons et al. [1] estimated the mortality risks for the transportation of chlorine by rail in the eastern part of the United States. It is estimated that once every ten years a damaged tank car will rupture, releasing its entire contents to the environment. The product of the area of the lethal plume and the population density of the area gives the number of potential mortalities. The probability of the occurrence of a given number of mortalities is the product of the accident frequency (0.1 per year), the probability of weather conditions at the time of the accident, and the probability of the accident occurring in an area of a given population density.

Assuming that the population densities along rail routes in the state of Ohio are typical of most areas in the eastern part of the United States and by assessing atmospheric stabilities and wind speeds for a number of locations in the eastern states, Simmons et al. plotted a histogram of potential mortalities, shown in Figure 3-1. On the basis of these predictions, there is about one chance in 100 that from 50 to 100 persons will die as the result of a rail accident involving the spillage of chlorine within the next year.

At an ambient temperature of 21.1° C the vapor pressure of chlorine is approximately 689 kPa (100 psig). When spilled, liquid chlorine immediately vaporizes cooling to -34.4° C where the vapor pressure reaches one atmosphere. The result is an initial highly concentrated cloud of chlorine, followed by a lower, steady release of vapor depending on the rate of heat absorption from the surroundings. A pool of liquid chlorine at -34.4° C will vaporize at a slower rate on land than in water where the rate of heat exchange is accel-

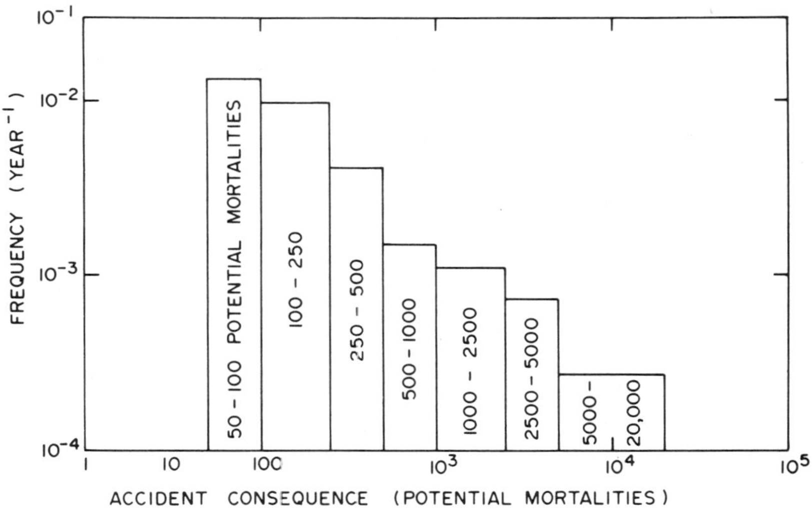

Figure 3-1 Accident frequency histogram. Rail tank car spills of chlorine, eastern United States. Source: J. A. Simmons et al. [1], Risk Assessment of Large Spills of Toxic Materials, Proc. 1974 National Conf. Control of Hazardous Material Spills, San Francisco, Aug. 25-28, 1974, p. 173.

erated. If a tank car containing 90 tons of chlorine at 21.1° C were to rupture, spilling its entire contents, about 17.5% (14.2 metric tons) would immediately flash while the remaining chlorine would spread out over the ground. For a pool on level ground that is 7.5 cm deep, the steady rate of evaporation in sunlight is about 27.3 kg/m²/hr. Adiabatic flashing and evaporation data for chlorine are shown in Figure 3-2.

The calculation of lethal plume area involves using the Pasquill-Gifford equation for ground-level concentration isopleths. Dosage is given by (see page 39):

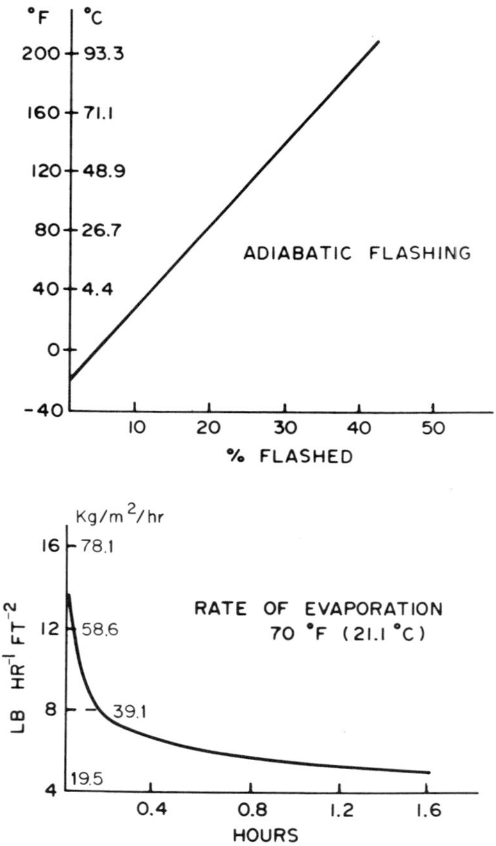

Figure 3-2 Evaporation data for liquid chlorine.

$$\text{Dosage} = \frac{Q}{\pi\sigma_y\sigma_z\bar{u}} \exp\left[-\frac{1}{2}\left(\frac{y}{\sigma_y}\right)^2 \right]$$

where Q = mass of vapor released in ppm

σ_y = standard deviation of the dispersion concentration in the crosswind direction

σ_z = vertical concentration distribution

$\bar{u}$ = average wind velocity and

y = crosswind distance at the center line of the dispersion

Based on the standard deviation values supplied by Turner [2] for various atmospheric stability categories, and considering the LD_{50} dosage for chlorine to be 1000 ppm/min, the area of lethal dose for a 90-ton tank car at 21.1° C is shown in Figure 3-3. At a wind

BASED ON TLV	OBSERVED AREA OF A SPILL, m^2 18.6	37.2	55.7	74.3
A. DOWNWIND, Km	1.3	2.1	2.6	3.1
B. CROSSWIND, Km	0.8	1.3	1.6	1.9
C. CIRCLE, m	146	219	274	311

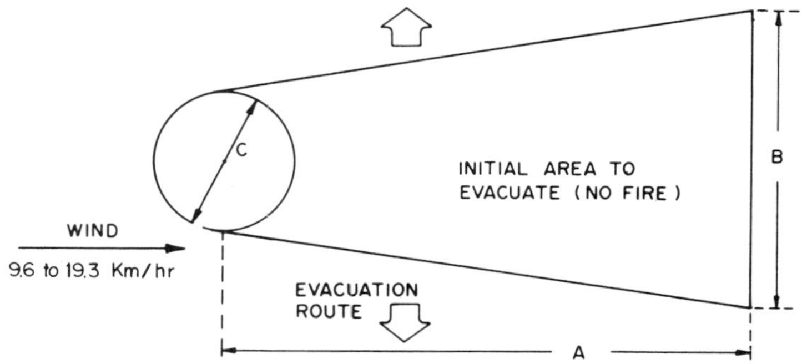

Figure 3-3 Area of lethal dose, initial flash vaporization from a 90-ton chlorine spill (14.2 metric tons initially flashed at 21.1° C. Source: J. A. Simmons et al. [1], Risk Assessment of Large Spills of Toxic Materials, Proc. 1974 National Conf. Control of Hazardous Material Spills, San Francisco, August 25-28, 1974, p. 171.

speed of about 2 m/sec (7.2 km/hr), the area of lethal dose at atmospheric stability D is about 6 km^2. The bar gives the area range for initial chlorine temperatures from 4.4 to 38° C.

The U.S. Department of Transportation [3] has developed an emergency guide for gases and highly volatile liquids that are toxic and/or extremely flammable, and shipped in bulk. This guide provides information on distances to evacuate in the event of spills. Information on evacuation for chlorine is shown in Figure 3-4. The degree of evacuation depends on the area of the spill, wind velocity, and the TLV of the hazardous material. For a spill of chlorine that covers an area of 74.3 m^2, in the presence of wind at 10-20 km/hr, the evacuation zone should extend 3 km downwind, 2 km across wind, and 155 m upwind of the spill site. It is cautioned that fragments of an exploding tank car in a fire can be hurled 609.6 m. Table 3-1 presents evacuation information on other highly hazardous materials

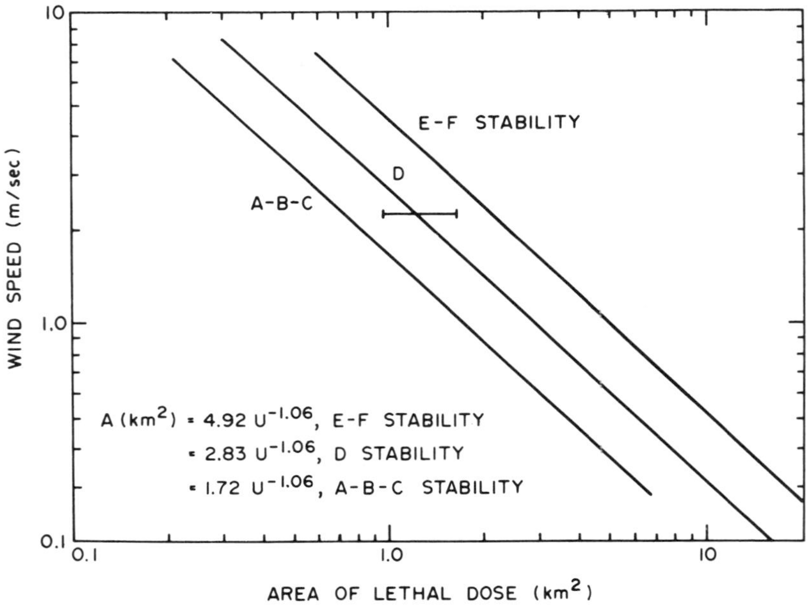

Figure 3-4 Evacuation distances for a spill of chlorine. Source: U.S. Dept. of Transportation, Emergency Services Guide for Selected Hazardous Materials, Spills, Fire, Evacuation Area, Office of the Secretary of Transportation, Washington, 1973.

Table 3-1

Initial Evacuation Distances for Various Highly Hazardous and High-Volume Materials

Material	Downwind, km (A)	Crosswind, km (B)	Circle, m (D)
Acrolein	7.6	4.5	695
Acrylonitrile	0.3	0.2	27.4
Ammonia	0.6	0.5	82.3
Carbon disulfide	0.3	0.2	41.1
Chlorine	2.1	1.9	310.9
Ethylenimine	3.4	2.1	347.5
Ethylene oxide	0.3	0.2	41.1
Fluorine	1.8	1.1	201.2
Hydrogen chloride	2.3	1.4	237.7
Hydrogen cyanide	1.1	0.6	118.9
Hydrogen fluoride	2.9	1.8	297.2
Hydrogen sulfide	1.3	0.8	146.3
Methyl amines	1.1	0.8	137.2
Methyl mercaptan	0.8	0.5	91.4
Nitric acid (fuming)	1.1	0.6	128.0
Nitrogen tetroxide	1.1	0.8	137.2

Source: U.S. Dept. of Transportation, Emergency Services Guide for Selected Hazardous Materials, Spills, Fire, Evacuation Area, Office of the Secretary of Transportation, Washington, 1973.
Note: Spill area 74.3 m^2; 9.6-19.3 km/hr wind; no fire, based on TLV.

assigned a 74.3 m^2 spill area in a wind of 10-20 km/hr. Readers who are interested in the details involved in the calculation of evacuation areas are referred to Turner's work on atmospheric dispersion estimates [2].

In general, once a pollutant enters a waterway, it cannot be contained. If the material is insoluble, retainment booms and adsorption media can be used, but it may be impossible to retain soluble materials. Volatile compounds are also lost to the atmosphere. Usually most environmental damages occur immediately when a spill of hazardous materials reaches the aquatic environment. For this reason the major emphasis is on the water environment.

How toxic a substance is to an organism depends on the nature of the organism, the nature of the toxicant, environmental conditions, the concentration of the toxicant, and the length and nature of the exposure in the environment.

A toxic substance may produce damaging effects in animals by absorption through external membranes, the ingestion of contaminated solids and fluids, or the inhalation of contaminated air. The result may be death or injuries such as burns, impairment of neurological function, reductions in physiological processes, etc. Lethal effects can result after the direct contact of animals with the substance or through accompanying changes in the environment involving pH, dissolved oxygen, toxicity, or turbidity. Marine organisms die because they cannot adapt to rapid environmental changes. Sometimes a combination of environmental effects is deadly. For example, the toxicity of nickel cyanide to fish is increased a thousandfold by a decrease in pH from 8.0 to 6.5 [4]. Turbidity can lower the concentration of dissolved oxygen in water and retard the beneficial actions of photobacteria. In addition, materials producing turbidity might settle out and smother benthic flora and fauna.

The processes that influence the fate of pollutants in the aquatic environment are outlined in Figure 3-5. Natural processes include physical distribution mechanisms involving dispersion and water currents, biological and chemical transformations, the biological concentration of pollutants, and various physical and chemical processes including adsorption, precipitation, and ion exchange. Pollutants can be widely distributed and selectively concentrated in the aquatic environment.

Certain microorganisms deactivate toxic substances by natural metabolic processes provided that concentrations of the pollutants

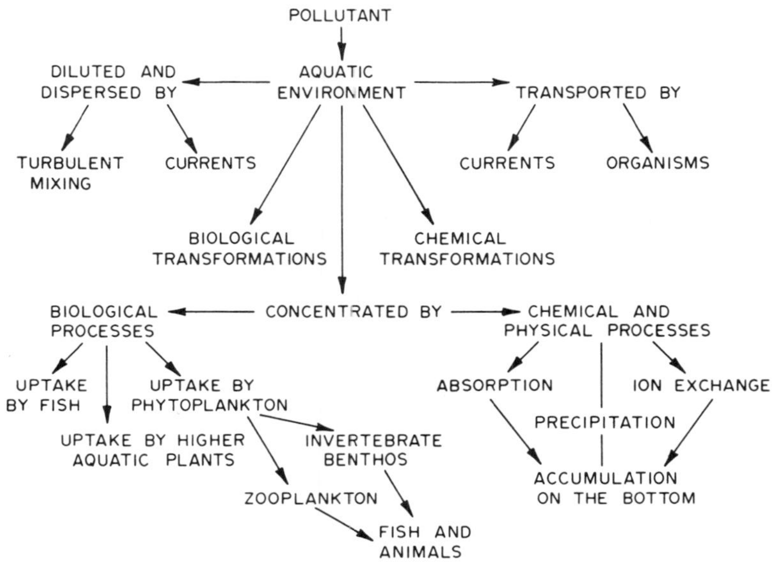

Figure 3-5 Environmental distribution processes affecting pollutants. Source: G. W. Dawson, A. J. Shuckrow, and W. H. Swift, Control of Spillage of Hazardous Polluting Substances, U.S. Federal Water Quality Administration, Program No. 1509, PF-7, Washington, 1970, p. 55.

do not exceed certain levels, although certain processes are undesirable. For example, sulfates being reduced to sulfides by microorganisms can produce odor problems and insoluble salts of mercury can be transformed by benthic organisms into soluble methyl mercury compounds that are highly toxic to animals. The rate of microbial transformations depends on the predominant flora and fauna of the environment, the type of available energy and nutrient sources, and ambient factors such as aeration, temperature, pH, and light. The processes that occur in the aquatic environment are varied and complex. Spills of highly hazardous substances produce shocks that quickly alter natural processes with potential disastrous consequences.

3-1 PERMISSIBLE ENVIRONMENTAL LEVELS OF HIGHLY HAZARDOUS SUBSTANCES

Permissible levels of hazardous substances in Ontario and U.S. public water supplies are presented in Table 3-2. The levels are identical with the exception of boron which is five times higher for the U.S. requirement. However, the recommended level is 1.0 mg/liter, which is the same as the Ontario level.

In 1973 the United States proposed effluent standards for highly toxic substances [5] which are outlined in Table 3-3. Proposed discharge rates depend on the type of toxicant and the nature of the receiving body of water and its assimilative capacity. The standards depend on whether the receiving water is fresh or salty, on the volume of the receiving water compared to the rate of discharge, and on the ability of the receiving water to disperse the pollutant. The standards reflect the TLm values and criteria mentioned earlier (Tables 2-1 and 2-5). A maximum has been placed on the total daily discharge of specified pollutants as well as on the rate of discharge. For example, 20 μg/liter of mercury is tolerated in effluents entering receiving waters where receiving water flow exceeds 0.283 m^3/sec or the receiving water is a lake greater in area than 202.3 ha. If a specified level of diffusion is not assured, the tolerated value is 2 μg/liter. No mercury is tolerated in effluents to receiving waters having a flow rate below 0.283 m^3/sec or to lakes having an area less than 202.3 ha. The maximum total daily discharge of mercury to any stream is 0.735 kg, any lake 0.612 kg, any estuary 12.2 kg, and any coastal water, 14.7 kg. Tolerated discharge rates to receiving bodies of salt water are five times higher than that of fresh water, depending upon the ease of diffusion.

The degree of damage to the aquatic environment following a spill relates to the toxicity of the pollutant as indicated by its LD_{50} or TLm values and to the residence time and volume of the receiving body of water. It is obvious that high concentrations of

Table 3-2

Permissible **Levels** of Hazardous Materials in Public Water Supplies, mg/liter

Materials	Ontario[a]	United States[b]
Inorganic		
Arsenic	0.05	0.05
Boron	1.0	5.0
Cadmium	0.01	0.01
Chromium	0.05	0.05
Copper	1.0	
Fluoride	1.7	1.4-2.4
Lead	0.05	0.05
Selenium	0.01	0.01
pH	6.0-8.5	
Organic		
Cyanide	0.20	0.20
Pesticides		
Aldrin	0.017	
DDT	0.042	
Dieldrin	0.017	
Endrin	0.001	
Phenol	Absent	<0.001

[a]Source: Ontario Water Management Guidelines and Criteria in Ontario, Toronto, Ontario, 1970, pp. 21-22.

[b]Source: G. L. Waldbott, Health Effects of Environmental Pollutants, Mosby, St. Louis, 1973. (From U.S. Dept. of Health, Education and Welfare, Public Health Service, Environmental Health Service, Bureau of Water Hygiene, Data-1970.)

toxicants will result when spills occur in slow-moving streams and small lakes.

Table 3-3

United States Effluent Standards for Highly Toxic Substances

Substance	Low flow[a]	Maximum Daily Discharge, mg/liter: Higher Flow: Immediate diffusion[b] Fresh	Immediate diffusion[b] Salt	Diffusion less than 1:10[c] Fresh	Diffusion less than 1:10[c] Salt	Maximum Total Daily Discharge, mg: Stream	Lake	Estuary	Coastal water
Aldrin-Dieldrin	0	0.5	5.5	0.05	0.55	0.0735	0.061	0.686	0.808
Cadmium (metal)	0	40	320	0.4	32	5.88	4.90	39.2	46.5
Cyanide	0	100	100	10	10				
DDT	0	0.2	0.6	0.02	0.06	0.0294	0.054	0.0234	0.088
Endrin	0	0.2	0.6	0.02	0.06	0.0294	0.054	0.0234	0.088
Mercury (metal)	0	20	100	2	10	0.735	0.612	12.2	14.7
PCBs	0	280	10	28	1	0.0294	0.0245	0.208	0.250
Toxaphene	0	1.0	1.0	0.1	0.1	0.1471	0.122	0.122	0.147

Source: Environmental Protection Agency, Water Programs, Proposed Toxic Pollutant Effluent Standards, Federal Register, 38: 247, Dec. 27, 1973.

[a]Discharge into streams, lakes, or estuaries with a low flow less than or equal to 0.283 m^3/sec or into lakes with an area of less than or equal to 202 ha.

[b]Discharge into streams, etc., with flow greater than 0.283 m^3/sec or into lakes with an area greater than 202 ha or into coastal waters having a flow greater than or equal to 10 times the waste flow with the provision that an immediate 1:10 diffusion is assured.

[c]Same as above[a], except that fresh water flow is less than 10 times the waste stream flow.

BIOLOGICAL EFFECTS

The effects of spills frequently extend beyond the initial destruction of life. Even after toxic substances have become diluted to a point where ambient water concentrations are below the thresholds of toxicity, undesirable effects can occur. A model for the concentration of DDT in the environment is presented in Figure 3-6.

After reaching the water environment, DDT is adsorbed by aquatic plants and fish, which in turn provide food for other animals such as fish-eating birds. During the process of going from water to birds, DDT may be concentrated one million times, i.e., by biomagnification. This process occurs for other chlorinated hydrocarbons in water and organometallic compounds such as methyl mercury.

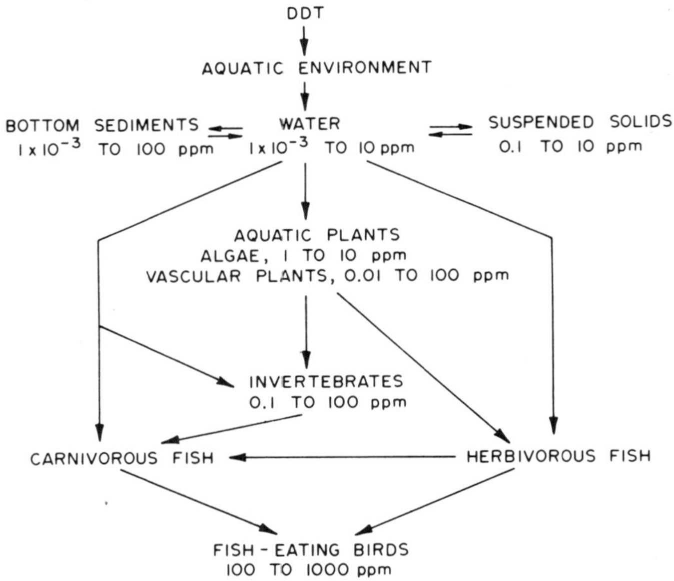

Figure 3-6 Model of DDT concentration and biomagnification in the aquatic environment. Source: G. W. Dawson, A. J. Shuckrow, and W. H. Swift, Control of Spillage of Hazardous Polluting Substances, U.S. Federal Water Quality Administration, Program No. 1509, PF-7, Washington, 1970.

Pyle [6] reports a case involving the concentration of DDT along food chains in Lake Michigan. A sample of sediment from the lake showed 8.5 ppb DDT. Invertebrates from the same area contained about 5 ppm DDT, and herring gulls tested at 3200 ppm in fatty tissues.

It is interesting to note that the permissible limit of DDT in water in Canada is 0.042 mg/liter (42 ppb) and the TLm is 2.3 ppb. According to Pyle's study, these levels can be readily concentrated by aquatic and land creatures. Humans contain 5-20 ppm DDT in fatty tissues, whereas food contains 0.05-0.5 ppm. The daily ingestion of the average person is about 0.1-0.2 mg [7].

The solubility of DDT in water at 30° C is only 10^{-6} g/liter according to Frear [8]. Nonpolar solvents such as kerosene will dissolve 100 g of DDT per liter, whereas benzene dissolves 780 g per liter. Animal fats are relatively low in polarity and tend to selectively absorb DDT from the environment with the greatest concentrations accumulating in the liver, brain, and gonadal tissues.

Factors influencing the persistence of pesticides, herbicides, and other toxic substances in the soil [9] are outlined in Table 3-4. The persistence of highly hazardous substances in the soil is regulated by the rate of leaching, adsorption, volatilization, and decomposition. Decomposition can involve microbial, chemical, and photochemical processes. The process of leaching of pollutants from soils may contaminate streams for long periods after a spill has occurred. Plastic sheets are frequently used to line sumps that have been dug to retain spilled substances to diminish soil contamination. Soil that has been contaminated by toxic substances must be excavated and removed to a disposal site.

The persistence times of some common pesticides in soil are presented in Table 3-5. These data are based on research by Lichtenstein and Schulz [10,11]. Seven pesticides were applied at a rate of 5.9 kg/ha to soil and worked 12.5 cm into the soil; samples were analyzed over a 5-month period. The data show that aldrin, a chlorinated hydrocarbon, is an extremely recalcitrant substance, whereas certain organophosphorus substances such as malathion disappear at a much faster rate.

Table 3-4

Factors Affecting the Persistence of Toxic Substances in Soil

Factor	Determined by
Leaching	Water solubility
	Soil permeability and texture
	Microbial and plant activity
	Amount and intensity of rainfall
Adsorption	Properties of the substance, organic and clay composition
	Type of soil
Volatilization	Depends on nature of the substance and the absorptivity of the soil
Decomposition	
Microbial	Temperature, pH, moisture, aeration, soil nutrients
Chemical	Oxidation, hydrolysis, complexing
Photochemical	Possibly important in arid, sunny areas

Source: G. Sykes and F. A. Skinner, Microbial Aspects of Pollution, Academic Press, New York, 1971, p. 235.

Temperature affects the rate of disappearance of hazardous substances from the soil. In one experiment [12] 92% of the original aldrin concentration remained in a soil specimen after four weeks at 7° C, whereas after incubation of a similar soil specimen at 46° C, 40% of the aldrin remained. Other factors that affect the rate of disappearance of toxic materials in soil include the nature of the crops, the cultivation and moisture content of the soil, and its susceptibility to microbial degradation.

Cripps [13] reports that certain molecular arrangements restrict the degradation of a compound by microorganisms. Microbial attack is retarded by the presence of an ether linkage, chlorine atoms, branched chains, and substituted amino groups. These groups are found in pesticide molecules and other industrial chemicals. The presence of two or more substituents in a benzene ring and the actual

Table 3-5

Persistence of Some Common Pesticides in Soil

Pesticide	Type	Half-Life	Time to Reach 0.1 ppm[a]
Aldrin	Chlorinated HC	2 months	
Carbaryl (Sevin)	Carbamate	1 month	
Phorate (thimet)	Organo P	1 month	
Azinphosmethyl (Guthion)	Organo P	20 days	
Parathion	Organo P	20 days	90 days
Methyl parathion	Organo P		30 days
Malathion	Organo P		8 days

Source: E. P. Lichtenstein and K. R. Schulz, Effects of Moisture and Microorganisms on the Persistence and Metabolism of Some Organophosphorus Insecticides in Soils with Special Emphasis on Parathion, J. Econ. Entomol., 57: 618, 1964. E. P. Lichtenstein and K. R. Schulz, Residues of Aldrin and Heptachlor in Soils and Their Translocation into Various Crops, J. Agr. Food Chem., 13: 57, 1965.

[a]3% of applied dose.

position of the substituents affect degradability. Meta derivatives are more recalcitrant than ortho and para derivatives. Multiple substitutions are even more resistant since there is increased reduction in the points on the ring where microbes can attack. In addition, certain structural arrangements produce steric effects preventing the necessary close associations for enzymatic action.

In 1970 about 30,840 metric tons of polychlorinated biphenyls (PCBs) were sold in the United States. PCBs are widely used in industry and are freely transported around the country. They are used in the preparation of plasticizers, heat-exchanger fluids, solvents, adhesives, hydraulic fluids, and sealants. About 907-1814 metric tons yearly escape into the atmosphere from plastics, and concentrations of PCBs greater than DDT and DDE have been found in arctic bears [14].

Different species of fish vary in their susceptibility to PCBs. Catfish and bluegills are reported to succumb to water concentrations of 20-50 ppb, trout to 8 ppb, and shrimp to as little as 1 ppb [15]. In birds, concentrations up to 1000 ppm have been found [16].

PCBs are similar to DDT with regard to the low solubility in water and tendency to accumulate in body fats. Liver tissue and body cells are inhibited in growth following absorption since PCBs inhibit enzymatic actions. PCBs are reported to enhance the toxic action of organophosphate pesticides and DDT when they are present in the environment together with these substances [17].

Certain metal cations such as Ag^{+}, Cd^{2+}, Th^{2+}, Pb^{2+}, Hg^{2+}, and Be^{2+} produce toxic effects in human beings and animals at very low concentrations. Perhaps evolutionary processes have caused animals to have a very low tolerance to these metals since they are present in the ecosphere in areas of low population density at very low concentrations. The main effect produced by some metals is that they coordinate with ligands to catalyze or retard biochemical reactions.

An increase of lead in the environment could result from a spill of tetraethyl lead (TEL) or some other lead compound. The toxicity of lead is caused by its inhibition of the synthesis of porphyrin-type compounds which contribute to oxygen transport in higher animals. The enzyme aminolevulinic acid dehydrase, which is required in the synthesis of hemoglobin, is inhibited when the lead concentration in the blood of animals exceeds a certain level. The American Conference of Governmental Industrial Hygienists [18] reports the threshold limit value (skin) of lead to be 0.1 mg/m^3. For man, a TEL concentration of 100 $mg(Pb)/m^3$ in air for 1 hr may produce illness, and a level of 60 μg of lead per 100 ml of blood is considered to mark the lower limit of lead poisoning [19]. Small children are very susceptible to lead poisoning, the lead accumulating in bony tissue where it retards the normal maturation of red blood cells leading to various degrees of mental retardation in the individuals affected [20,21].

In the late 1960s, mercury came under investigation as an environmental toxicant. Jensen and Jernelöv [22], in 1969, were among the first to suggest that bacteria have the capacity to transform inorganic mercury into organic form readily adsorbed by animals. It was proposed that inorganic mercury was methylated to methyl and dimethyl mercury by methylating bacteria. This hypothesis was proved using lake sediment impregnated with mercuric chloride.

The process of methylation is not yet fully understood. In one process a methyl group coordinated to a cobalt ion is transferred to an Hg^{2+} ion. In another, the enzyme methionine synthetase and a coenzyme methylcobalamin together methylate mercury. Apparently, many microorganisms possess this capacity [6].

Suggested biotransformations for mercury in the aquatic environment are outlined in Figure 3-7. Inorganic mercury and industrial compounds, such as the fungicides panogen and ceresan, enter water and accumulate in bottom sediments where bacterial transformations occur. The methyl and dimethyl mercury compounds released from the sediments are absorbed by fish or returned to the atmosphere. The distressing fact that mercury is reported to return to the atmosphere

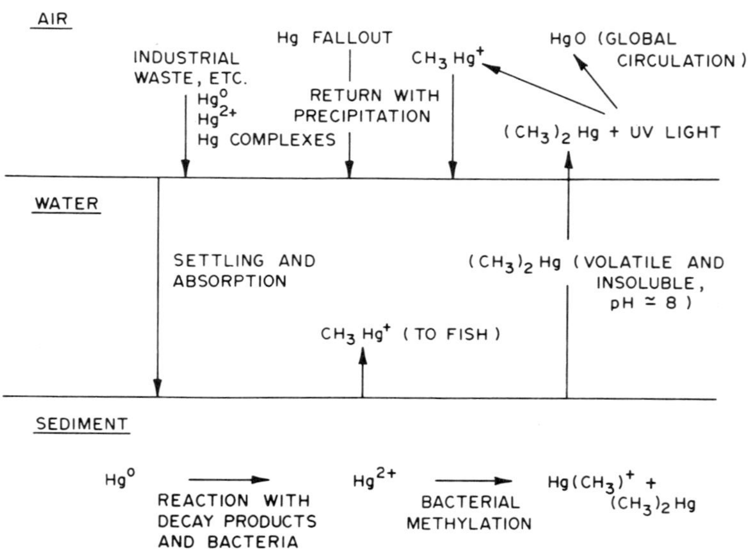

Figure 3-7 Biotransformations of mercury. Source: D. G. Langley, Mercury in an Aquatic Environment, JWPCF, 45(1): 14-51, 1973.

suggests far-reaching and long-lasting effects with the involvement of a complex biogeochemical cycle in the biosphere.

Fish have the ability to concentrate mercury through the ingestion of contaminated food and by the absorption of methyl mercury. In one experiment trout were exposed for 1 hr per day to a 0.1 ppm concentration of ethyl mercury [23]. After ten days, the concentration of mercury in the muscle protein of the fish reached 4 ppm. The ability of mercury to concentrate in animals can be compared to the model presented for DDT with mercury accumulating in the protein substances of animals.

Methyl mercury is absorbed through the skin, respiratory tracts, and alimentary canals of animals with high efficiency. After entry into the body, methyl mercury prefers liver, kidney, and brain tissues and accumulates in red blood cells. In mammals it passes through the placenta and reaches the fetus, where it concentrates at levels above those in the organs of the mother. Cases of mental retardation in the regions of Minamata and Niigata Japan are well known.

Very little is known about the effects of methyl mercury at the molecular level. It is known that methyl mercury is bound to proteins, but the specific proteins affected have been examined only superficially. It is known that mercury reacts with sulfhydryl (SH) groups which are present in certain natural amino acids.

Flynn [24] reports that the carp harvested from Lake Erie near Cleveland contained concentrated amounts of trace metals. Clippings from the pelvic fins of carp were analyzed for cadmium, copper, lead, iron, magnesium, zinc, and manganese along with water samples from the areas where the clippings were taken. Eighteen samples were taken within 4.8 km of the effluent from the Cuyahoga river and 20 samples at an area 48 km west of this location. Data on the levels of metal in the water are listed in Table 3-6 along with Ohio-EPA standards. Concentration factors for the fish are also shown where concentration factor is defined by:

Table 3-6

Metal Concentration Levels in Lake Erie Water and Fish, ppm

Metal	Ambient Water Levels 4.8 km[a]	48.3 km[a]	Ohio-EPA Standards
Cadmium	0.085	0.018	0.005
Copper	0.14	0.22	0.5
Iron	1.13	0.79	1.0
Lead	0.11	0.05	0.04
Magnesium	3.65	3.54	--
Manganese	1.06	0.93	1.0
Zinc	1.28	0.88	1.0
	Concentration Factors for Fish		
Cadmium	124.7	508.3	
Copper	48.6	32.7	
Iron	62.0	91.1	
Lead	55.0	102.0	
Magnesium	16.4	21.2	
Manganese	4.6	5.4	
Zinc	101.2	136.4	

Source: A. Flynn, Biological Availability of Ions in Heavy Metal Spill, Proc. 1974 National Conf. Control of Hazardous Spills, San Francisco, August 25-28, 1974, p. 260.

[a]Distance from Cuyahoga river.

$$CF = \frac{C_f}{C_w}$$

where C_f and C_w are metal concentrations in fish and water, respectively.

Concentrations closely reflect the background levels of the trace metals in the water.

Phenol is used extensively in some refineries as a selective solvent for the refining of lube oil distillates and, less extensively, as a base chemical and germicide. It is generally handled in closed systems, but contact with this substance is permitted in cases of sudden breaks and spills. On contact with the skin, phenol or phenolic solutions produce severe burns. The extent of absorption of phenol through skin appears to be determined mainly by the area of the exposed skin rather than the concentration of the solution [25]. Phenol is classified as a general protoplasmic poison which acts on tissue by denaturing and coagulating proteins.

In aqueous solution phenol is classed as a taste- and odor-producing compound [26]. Chlorination of phenol-containing water forms chlorinated phenols, which are far more toxic. Canadian public waters are required to be totally free of phenol, while up to 0.001 mg/liter is tolerated in U.S. water. The TLm value for fish is 0.1 mg/liter. One of the most effective ways of removing phenol from water is by carbon adsorption, which is discussed in Section 6-2.5. Cresol is also a general protoplasmic poison, similar in action to phenol, but less toxic. The meta form of cresol is the least toxic [27].

Among the most toxic liquids handled in industry and transported by rail and truck are liquid industrial wastes. Almost any kind of highly hazardous substance can be found in industrial wastes, from organic substances such as phenol to inorganic salts of mercury, chromium, cadmium, etc. A highway accident releasing quantities of such materials could cause considerable damage to the environment. A summary of industries and toxic wastes associated with these industries is presented in Table 3-7. The frequency of occurrence of such materials as phenol, heavy metals, chlorine, cyanides, and oil emulsions in industrial wastes should be noted. Chlorine and cyanides produce fish kills at very low concentrations in water. The TLm value for chlorine is 0.03 mg/liter and for CN^- (HCN) it is 0.05 mg/liter. A detailed account of a cyanide spill is presented in Section 3-3.1.

Table 3-7

Toxic Liquid Industrial Wastes

Industry	Toxic Components in Waste
Pesticides	Chlorinated hydrocarbons
Caustic production	Mercury and dissolved chlorine (mercury cells have in most cases been replaced by diaphragm cells)
Sodium chlorate production	Free chlorine and free chlorine dioxide and chromium
Glass	Fluorides and oil emulsions
Petroleum	Phenols, sulfides, mercaptans, and traces of heavy metals and fluorides
Chemical	Phenols, cyanides, acrylonitrile, phosphorus, and heavy metals
Iron and steel	Phenols, cyanides, fluorides, pickling solutions, chromium, zinc, and tin
Pulp and paper	Phenols, mercaptans, mercury, chlorine, and essential oils
Mining (nonferrous metals, excluding uranium)	Copper, lead, zinc, mercury, cadmium cyanides, and traces of heavy metals
Copper	Heavy metals
Lead	Heavy metals, arsenic
Metal finishing	Cyanide, chromium, heavy metals, oil emulsions
Automotive	Lead, chromate, phenols, heavy metals, oil emulsions
Leather (chrome tanning)	Trivalent chromates, sulfides
Thermal power (ash lagoon)	Heavy metals including mercury
Phosphoric acid (thermal process)	Colloidal elemental phosphorus

Source: Canada Department of Fisheries and Forestry Study of Special Liquid Industrial Wastes, Montreal Engineering Co. Ltd., Montreal, 1971, Parts 1-5 and 6.

The effect of industrial chemicals on municipal waste water treatment plants is being assessed by the Alleghany County Sanitary Authority [28]. The assessment involves a survey of nearly 5000 industries in the Pittsburgh area to identify what hazardous materials were used in the area and to carry out pilot plant studies. Immediate studies were conducted on cadmium chloride, sulfuric acid, sodium hydroxide, phenol, and methanol. Additional studies were scheduled for ammonium chloride, 2,4-D, sodium cyanide, acrylic acid, chromium, and tetraethyl lead. Sulfides and sulfites are often overlooked as toxic wastes.

Initial results reveal that concentrations of 500 mg/liter of cadmium chloride in the influents to the pilot sewage treatment plant seriously affected the BOD (biological oxygen demand) removal efficiency, whereas influent containing 100 mg/liter had no discernible effect. The plant was found to recover fairly rapidly from a sulfuric acid spill within 12 hr. During the spill, the influent reached a pH of 2 for 2 hr.

Hydrogen sulfide in natural and petroleum gases is removed by oxidizing the sulfurous gas to elemental sulfur with air. The Province of Alberta in Canada has accumulated a vast amount of sulfur (approximately 9 million metric tons). About one-half of the sulfur produced in Alberta is converted to sulfuric acid or fertilizer and the remainder is stockpiled in the open.

Sulfur is usually transported in the form of "slate" or pellets. During transportation, it may become powdered presenting an explosion hazard, or, if spilled into a watercourse, microbes may oxidize the sulfur with the formation of sulfuric acid.

With new uses for sulfur developing at a great rate, much larger quantities may be transported in the future, increasing the expectancy of spills.

3-3 SELECTED CASE HISTORIES OF SPILLS

3-3.1 Highly Toxic Substances

The Endrin Spill at Shawnee Lake, Ohio [29]

On June 2, 1971, a man who was denied permission to swim at Shawnee Lake, Ohio, threw a solution of about 700 g endrin mixed with strychnine-treated corn into Shawnee Lake. The entire fish population of the lake was destroyed with the exception of tadpoles. Shawnee Lake is about 304.8 m long, 76.2 m wide at its widest point, and, on the average, about 4.6 m deep. Endrin concentrations exceeded 7 ppb in the water of the lake and 100 ppb in the sediments along the banks of the lake.

Endrin is extremely poisonous to fish even at concentrations as low as 0.2 ppb. The safe limit for humans is 1.0 ppb. The half-life of endrin ranges from seven to twelve years and it is capable of concentrating up to 10,000 times in the environment.

Immediate steps were taken to isolate the lake from the environment and to recover the endrin by immersing bags of activated carbon in the effluent spillway from the lake and pumping the effluent back into the lake. A filter was constructed using a 1.2 x 2.4 x 5.5-m wooden box containing coarse gravel covered with a mesh screen over which a 2-m deep bed of charcoal was placed. Water was pumped from the lake through this filter at a rate of about 4.85 liters/sec. Unfortunately, underground springs fed the lake at a higher rate than the pumping rate.

The next technique used was a large filter composed of 160 bales of hay through which water was pumped at 90.9 liters/sec. No detectable endrin was found in the water from the hay filter and the lake was pumped dry.

About 4970 m^3 of sediment was next removed from the bottom and sides of the lake and transferred to a disposal area where the sediment was mixed with clay to achieve 1 ppb of endrin in the diluted soil.

On September 1, 1971, the refilled lake was restocked with fish. The total cost of the cleanup was $300,000.

Cyanide Pollution at Dunreith, Indiana [30]

On January 1, 1968, a train wreck occurred at the town of Dunreith, Indiana, and spilled 4.37 m^3 of chemicals containing 1270 kg of cyanide into Buck Creek. Five 72.8-m^3 tank cars were involved, two of which carried acetone cyanohydrin, one vinyl chloride, another ethylene oxide, and the last methyl methacrylate. Fire and explosions resulted in the production of toxic gases which prompted the evacuation of townspeople. Acetone cyanohydrin decomposed into acetone and the highly poisonous hydrogen cyanide.

The following day dead cattle and dead fish were observed along the banks of the creek. An investigation revealed seven dead cattle, several hogs, and thousands of dead fish. Water samples revealed the presence of 405 mg/liter of cyanide and an odor of hydrogen cyanide proving that the acetone cyanohydrin had undergone decomposition in the water.

On the fourth and fifth days after the spill (January fifth and sixth) a total of 2812 kg calcium hypochlorite was added slowly to the stream. The calcium hypochlorite converted the cyanide to cyanate. The treated water was tested using live minnows to assure that no toxicity remained. Treatment was maintained until after the snow had melted since cyanide from the contaminated soil continued to seep into the creek.

It is believed that immediate chlorination after the spill might have prevented the deaths of a large number of fish; however, the chlorine itself might have killed some fish.

Investigation following the incident indicated that a broken rail was responsible for the derailment that caused the train wreck. It is reported that the railroad company paid the cost of the hypochlorite treatments which amounted to $3083.

3-3.2 Toxic Gases

The Camrose Hydrogen Sulfide Emergency [31]

At about 3:45 p.m. on October 2, 1973, a Sun Oil Company oil well near Camrose, Alberta, erupted, releasing natural gas and hydrogen sulfide into the atmosphere. The well spewed gas for about 26 hr and this gas contained about 5% hydrogen sulfide. By 6:30 p.m. the escaping gas had accumulated as a large cloud advancing northeast at about 1.52 m/sec.

The Royal Canadian Mounted Police were immediately notified by the oil company of the eruption and area residents were ordered to evacuate. Radio CFCW of Camrose promptly broadcast information for the benefit of area listeners.

Hydrogen sulfide concentration in the atmosphere was measured by the environmental officials of Alberta and found to be on the order of 0.02 ppm. It was estimated that concentrations of H_2S reached 5-10 ppm in some areas, and one value of 50 ppm was reported 2.4 km from the wellhead.

By 9:00 p.m. the Emergency Government Headquarters of the City of Camrose was fully operational and more than 300 evacuees were processed over the next 4 hr. In the end it was estimated that a total of 800 persons left their homes in the disaster area.

The well was successfully capped by 6:00 p.m. the next day and people returned to their homes. No personal or animal injuries were reported. The Sun Oil Company assumed financial responsibility for the costs incurred for emergency operations, including evacuee reception and care. Provincial Government costs were charged against the Emergency Contingency Fund.

The Sarnia Chlorine Emergency

In August of 1973 in Sarnia, Ontario, an undisclosed amount of chlorine was released to the atmosphere by the Dow Chemical Company of Canada causing 200 construction workers to be hospitalized. The workers were located downwind at the Sun Oil Company about 0.4-0.8 km away. They were enveloped in the cloud of vapors that drifted over the work area.

The accident was reportedly caused by human error when a maintenance worker removed by mistake the flanges holding the bonnet on a valve in a liquid chlorine line. Liquid chlorine escaped from the valve and rapidly vaporized after hitting the warm ground.

Most of the workers affected by the chlorine vapors were released from the hospital after overnight treatment and no deaths occurred. Vegetation was "burned" about 0.8 km downwind; however, these plants recovered subsequently.

Dow reported that action was taken within a few minutes after the onset of the leak when sensors responded to the presence of chlorine vapors. The sequence of actions were approximately as follows:

1. The leak was isolated by closing line valves upstream and downstream of the leak.
2. Dow's inplant emergency procedure was put into effect. Employees assembled at designated areas to await orders.
3. A water curtain was established around the leak and the escaping liquid was insulated with foam. Water curtains were achieved using high-pressure fog nozzles.
4. The Sarnia Emergency Operations Center, which is located at the City Police Headquarters, was notified of the emergency and the Chemical Valley Emergency Control Organization Plan was coordinated and placed into action. Incoming traffic to the spill area was halted and only emergency vehicles were permitted passage. Outward moving traffic was accelerated. Hospitals were notified regarding the nature of the emergency and the fire department stood by with respirators, loud-hailers, etc. Ambulances were directed to the area and industrial personnel assisted the police.
5. The leak was permitted to exhaust itself.

Officials of the Ontario Department of the Environment and the Dow Chemical Company reviewed the incident to weigh what actions might be taken to prevent a future incident of this type.

3-3.3 Spills of Corrosive Liquids

The Welland Sulfuric Acid Spill [32]

On December 1, 1972, a Toronto, Hamilton, and Buffalo train was derailed, dumping 1360 metric tons of 93% sulfuric acid near Welland, Ontario. Emergency proceddings were first initiated by police, firemen, and the works department crews. After news of the spill reached the Ontario Minister of the Environment, the Province of Ontario Contingency Plan for Spills of Oil and Other Hazardous Materials (see Section 4-1.1) went into effect.

The area was cordoned off by police and dikes were constructed to contain the flow of spilled acid. Most of the acid was contained within 0.8 km downgrade of the wreck. A number of families were evacuated from the area.

After about 6 hr, neutralizing materials were forwarded to the area by the team of the Ministry of the Environment. At approximately the same time, backhoes began to dig lagoons to contain the acid and drain the ditches. Although a quick thaw swelled the ditches and permitted some sulfuric acid to enter the Welland River, the lagoons held the bulk of the acid solution. Water wells in the area were not affected.

More than 1995 metric tons of soda ash and 9.07 metric tons of caustic soda were trucked to the spill area to neutralize the acid. Within two weeks, the acid was neutralized and the lagoons were pumped out along with the contents of the wrecked cars, and the acid was transferred to a disposal area. The total cost involved is not known.

The Sulfuric Acid Spill at Black Creek, Mississippi [33]

The consequences of the spill of sulfuric acid at Black Creek, Mississippi, were much more serious than those of the Welland spill.

On September 22, 1969, between 3000 and 5000 liters of concentrated sulfuric acid spilled into Black Creek, killing 350,000 fish. The pH of the water in the creek was reduced to 2.5. The majority of freshwater fish can tolerate a pH range from 6.5 to 8.4.

The Gulf Oil Company agreed to restock the stream and restore other damages at a cost to the company of approximately $20,650.

The Clinch River Fish Kill [34]

On June 10, 1967, a dike containing alkaline waste from fly ash for the Appalachian Power Company at Carbo, Virginia, collapsed and about 4 million m^3 of caustic solution was discharged into the Clinch River. As a result, an estimated 216,000 fish were killed in 145 km of river in both Virginia and Tennessee. All the food organisms in the 6-km stretch of river immediately below Carbo were completely eliminated; benthic and fish communities were not fully recovered two years after the accident.

Methods for neutralizing the spill were considered, but it was estimated that 453.5 metric tons of sulfuric acid would be required and this supply was not available. A series of meetings were held to decide how to avoid a similar disaster in the future. It was observed that the physical response to the incident was too slow and a more efficient reporting system was required. In addition, facilities should have existed for, at least, a certain degree of neutralization of the wastes after the spill. A study made of why the dike failed revealed little.

The Appalachian Power Company agreed to pay for damages, the sum of which has not been disclosed.

The Brandywine Creek Acid Spill, Pennsylvania

Over 25.5 m^3 of a solution containing 60% sulfuric acid and 40% nitric acid was spilled into Brandywine Creek near Downington, Pennsylvania, on February 6, 1973, as a result of the derailment of a Penn-Central train. Environmental Protection Agency personnel were called to the scene and the U.S. Coast Guard's National Strike Force was alerted. At least 2000 people were evacuated from the surrounding area and several persons were treated for the inhalation of corrosive vapors. Sodium carbonate was added to the water in an attempt to neutralize the acids and the downstream water intake of Wilmington, Delaware, was closed until the acid slug passed. Fortunately no serious injuries resulted from this spill, but there are no reports of the full extent of aquatic damages.

Tank Truck Spill of Sulfuric Acid, Unity, Ohio

On February 13, 1974, a highway tanker accident near Unity, Ohio, resulted in the spillage of 7.28 m^3 of sulfuric acid into a stream that feeds to the municipal water supply of Youngstown. EPA and state personnel supervised decontamination operations while the United States Coast Guard (USCG) remained on alert. A retaining dam was constructed downstream of the spill area and lime was added to the water to neutralize the acid. The treatment was considered effective since the Youngstown water supply showed no marked quality deterioration.

3-3.4 Chemical Spills

Rail Spill of Phenol, Slabtown, Maryland

A cargo shift on a curve caused a Western Maryland freight train to derail near Slabtown, Maryland, on June 27, 1972. Two tank cars ruptured, spilling 91 m^3 of phenol; about 6% of the total volume flowed into a nearby stream. Initially the concentration of phenol in the water reached about 1000 mg/liter and 0.05 mg/liter 16 km downstream. A collection channel was dug below the spill site to prevent rainfall from washing a slug of phenol from the soil into the stream. Effluent from the channel was directed through a rectangular box containing over 28.3 m^3 of granular activated carbon. The bottom layer in the box contained about 6.35 metric tons of 1.27- to 2.54-cm gravel to a depth of 30.5 cm, and was covered with a 7.62-cm layer of fine gravel. The carbon depth measured 3.05 m and provided an initial phenol reduction from 1000 to 0.5 mg/liter. By May 17, 1973, the remaining soil concentration of phenol had diminished sufficiently to permit bacterial decomposition to begin. The adsorption process was maintained until October 3, 1973, after which phenol influent concentrations were reduced sufficiently to remove the bed. The railway company bore the main costs of the cleanup.

Train Derailment, Chemicals Spill, Rush, Kentucky

On October 31, 1973. a 15-car train derailment occurred near Rush, Kentucky, resulting in the rupture of three tank cars containing acrylonitrile, metallic sodium, and other unidentified materials. Fortunately, a tank car of tetraethyl lead did not rupture. The area EPA team assisted by the USCG handled the emergency assisted by the EPA Technical Assistance Data System. About 100 area residents were evacuated and fortunately there were no casualties or uncontrollable fires or explosions. Area residents were allowed to return to their homes shortly afterward.

Leakage of PCBs from a Highway Transport, Kingston, Tennessee

Serious soil contamination resulted when polychlorinated biphenyls leaked from a transformer being transported by a highway carrier near Kingston, Tennessee, on March 7, 1973. The EPA Emergency Response Team for the area supervised cleanup operations while the USCG and other agencies stood on alert. Twenty-three truckloads of contaminated soil were removed from the spill area and transported to a disposal site in Texas. In spite of the cleanup proceedings, wells in the area showed varying degrees of contamination with PCBs. Water-treatment equipment has been installed at the spill site and monitoring of ground water will be required for five years with the total costs of cleanup and monitoring to be paid by the company responsible for the incident.

3-3.5 A Pipeline Spill

A Pipeline Failure and Crude Oil Spill in Wisconsin [35]

A rather dramatic pipeline failure occurred in Jefferson County, Wisconsin, on March 17, 1973. About 994 m^3 of crude oil were released in 40 min in a geyser up to 30.5 m in height after a flange gasket failed. The incident was first reported by a passing motorist. Immediate response by the EPA limited the travel of oil to 1.6 km and the total oil-soaked area to about 2.4 ha.

After the incident was reported, the pumping station at Cambridge, Wisconsin, was shut down and valves were closed, isolating the leak. The ruptured gasket had produced a 77.4-cm^2 opening for the oil to escape.

The escaped oil pooled in an adjacent cornfield and from there flowed into a drainage ditch and then to a cattail marsh. A sandbag-plastic sheet dam was constructed downstream from the marsh to prevent the oil from reaching a lake. Three 25.4-cm culverts at the base of the dam permitted water drainage while trapping the oil. Several barrier booms were placed along the drainage path to further trap the oil upstream of the dam. A 457-m road was constructed to permit pumpers and tank trucks to reach the contained oil pools.

Approximately 90% of the oil was recovered and returned to the pipeline. Unrecoverable oil remained as a scum adhering to vegetation. It was decided that the residual oil should be burned by using propane torches. This caused a voluminous discharge of clouds of black smoke which spread tarry residues over 1.6 additional hectares.

By May 15, only 8 days after the burn, a new growth of cattails had already appeared and by midsummer the vegetation had returned to normal. The toll of wildlife deaths was reportedly minor. There is a considerable body of literature on oil spills on land, in the ocean, and even in the ground.

A set of spill statistics for hazardous substances is presented in Appendix A. Ontario statistics are listed in Tables A-1 and A-2, and United States statistics in Table A-3.

Ontario statistics involve nonpetroleum substances that were spilled from trucks, rail cars, and storage/transfer facilities from January 1, 1971, to July 31, 1972. Table A-1 lists the spills classified as the type of material released and the source of the release, while Table A-2 summarizes the statistics in Table A-1.

The list of spills in Appendix A has been classified according to the categories of highly hazardous substances listed in Chapter 2. The data cover the period from June 1, 1967, to June 1970.

REFERENCES

1. J. A. Simmons et al., Risk Assessment of Large Spills of Toxic Materials, Proc. 1974 National Conf. Control of Hazardous Material Spills, San Francisco, August 25-28, 1974, p. 166.

2. B. D. Turner, Workbook of Atmospheric Dispersion Estimates, U.S. Dept. of Health, Education and Welfare, Publication P.B. 191 482, 1970. Distributed by the National Technical Information Service, U.S. Dept. of Commerce, Springfield, Va.

3. U.S. Dept. of Transportation, Emergency Services Guide for Selected Hazardous Materials, Spills, Fire, Evacuation Area, Office of the Secretary of Transportation, Washington, 1973.

4. R. A. Taft, Fish Toxicity and Physiology as Related to Water Pollution, Report 16, Sanitary Eng. Center, Cincinnati, Ohio, 1953.

5. Environmental Protection Agency, Water Programs, Proposed Toxic Pollutant Effluent Standards, Fed. Register, 38, No. 247, Dec. 27, 1973.

6. J. L. Pyle, Chemistry and the Technological Backlash, Prentice-Hall, Englewood Cliffs, N.J., 1974.

7. D. L. Dahlston, R. Garcia, and J. E. Laing, Pesticides, Scientists Institute for Public Information, New York, 1970.

8. D. E. H. Frear, Chemistry of Pesticides, D. Van Nostrand Co., New York, 1955.

9. G. Sykes and F. A. Skinner, Microbial Aspects of Pollution, Academic Press, New York, 1971.

10. E. P. Lichtenstein and K. R. Schulz, Effects of Moisture and Microorganisms on the Persistence and Metabolism of Some Organophosphorus Insecticides in Soils with Special Emphasis on Parathion, J. Econ. Entomol., 57: 618, 1964.

11. E. P. Lichtenstein and K. R. Schulz, Residues of Aldrin and Heptachlor in Soils and Their Translocation into Various Crops, J. Agr. Food Chem., 13: 57, 1965.

12. American Chemical Society, Cleaning Our Environment, The Chemical Basis for Action, Washington, 1969.

13. R. E. Cripps, The Microbial Breakdown of Pesticides, Shell Research Ltd., Borden Microbiological Lab., Sittingbourne, Kent, England, 1971.

14. G. L. Walbott, Health Effects of Environmental Pollutants, Mosby, St. Louis, 1973, pp. 225-230.

15. D. L. Stalling and F. L. Mayer, Jr., Toxicities of PCB to Fish and Environmental Residues, Environ. Health Persp., April, 1973.

16. A. G. Johnels, PCB: Occurrence in Swedish Wildlife, Natl. Swedish Environ. Protection Board, Solna Research Secretariat, Proc. PCB Conf., Stockholm, 1970, pp. 29-42.

17. M. Berlin, PCB, Effects on Mammals, Natl. Swedish Environ. Protection Board, Solna Research Secretariat, Proc. PCB Conf., Stockholm, 1970, pp. 44-50.

18.a. American Conference of Governmental Industrial Hygienists, Documentation of the Threshold Limit Values for Substances in Workroom Air, 3rd ed., 1971.

b. H. E. Christensen, ed., The Toxic Substance List, 1972 ed., U.S. Dept. of Health, Education, and Welfare, HSM 72-10265, Health Services and Mental Health Administration, Washington, 1972.

19. A. J. Flemming, Lead Symposium, Kettering Laboratory, Univ. Cincinnati, Feb. 25-27, 1963.

20. A. Tucker, The Toxic Metals, Ballantine Books, New York, 1972.

21. J. J. Chisolm, Lead Poisoning, Sci. Am., 224(2): 15, 1971.

22. S. Jensen and A. Jernelöv, Biological Methylation of Mercury by Aquatic Organisms, Nature, 223: 753-754, 1969.

23. A. L. Hammond, Mercury in the Environment: Natural and Human Factors, Science, 171: 788, 1971.

24. A. Flynn, Biological Availability of Ions in Heavy Metal Spill, Proc. 1974 National Conf. Control of Hazardous Material Spills, San Francisco, Aug. 25-28, 1974, pp. 257-261.

25. American Petroleum Institute, Toxicological Review, 1948, Phenol, API Dept. of Safety, New York, 1948.

26. MCA, Chemical Safety Data Sheet SD-4, Properties and Essential Information for the Safe Handling and Use of Phenol, MCA Inc., Washington, 1964, p. 18.

27. American Petroleum Institute, Toxicological Review, 1948, Cresol, API Dept. of Safety, New York, 1948.

28. A. P. Pajak et al., Management of Hazardous Material Spills in Municipal Wastewater Systems, Proc. 1974 Conf. Control of Hazardous Material Spills, San Francisco, August 25-28, 1974, pp. 58-64.

29. W. B. Nye, The Hazardous Material Spill Experience in Shawnee Lake, Ohio — A Case History, Ohio Dept. of Natural Resources, Proc. 1972 National Conf. Control of Hazardous Material Spills, Houston, Texas, March 21-23, 1972.

30. G. W. Dawson, A. J. Shuckrow, and W. H. Swift, Control of Spillage of Hazardous Polluting Substances, U.S. Federal Water Quality Administration, Program No. 1509, Contract No. 14-12-866, Washington, 1970, pp. 38-41.

31. Report on Camrose Area H_2S Emergency, 2-3 October, 1973, Alberta Disaster Services, Edmonton, Alberta, Nov. 15, 1973.

32. Environment Ontario Publication, Legacy, 2, No. 1 (Jan./Feb.), 1973.

33. U.S. House Document 92-70, 1971, Control of Hazardous Polluting Substances, A Report on Control of Hazardous Polluting Substances, Pursuant to Section 12(g) of the Federal Water Pollution Control Act as Amended, U.S. Govt. Printing Off., Washington, 1971.

34. Clinch River Fish Kill, U.S. Dept. of the Interior, FWPCA, Mid-Atl. Region, Charlotteville, Virginia, June, 1967.

35. R. O. Ostrander, Spill Response Planning and Operation in Wisconsin, Proc. 1974 National Conf. Control of Hazardous Material Spills, San Francisco, August 25-28, 1974, pp. 51-55.

Chapter 4

PLANNING FOR EMERGENCIES

In the early 1970s, intensive efforts were started to develop systems to meet environmental emergencies. Before this time, the emphasis was placed on the development of effective ways to identify hazardous substances, to define physical hazards associated with the handling of these substances, and develop emergency first aid procedures. Starting in 1970, increasing attention has been directed toward the environmental effects of pollutants and the development of procedures to prevent and cope with emergencies. In this chapter a number of systems will be discussed which have been designed to handle spills of hazardous substances.

The primary purpose of emergency plans is to prevent incidents as well as to initiate timely, well-organized responses. When developing an emergency plan for a given area, it is necessary to involve many agencies and personnel in the creation of a complex network of facilities that must function rapidly and precisely in the event of an emergency. It is of utmost importance that an incident be reported as soon as possible and that effective actions follow forthwith. The spilled substance has to be identified, appropriate personnel and equipment have to be rushed to the scene, and the cleanup has to be accomplished without endangering human life, while holding environmental damage to a minimum. It is becoming a standard practice for governments to order all handlers of hazardous material to design and to maintain emergency programs. Legislation is becoming increasingly demanding with regard to equipment designs for facilities handling hazardous substances. On preparing for emergencies, it is necessary to maintain community inventories with regard to personnel, equipment, and sources of hazardous substances. A working emergency plan will develop into a far-reaching system drawing on the technical resources of agencies in both the United States and Canada.

4-1 CONTINGENCY PLANNING

4-1.1 The Environment Canada Plan

In 1971, Environment Canada published a field manual [1] to assist in the development of countermeasures for spills of oil or toxic materials and to promote the concept of "preparedness." Most important in the plan are the regional coordinators who supervise teams of local on the scene coordinators (OSC). It is the function of coordinators to develop effective countermeasures for spills.
Canada has been divided into four main regions as shown in Figure 4-1. These regions are supervised by four regional coordinators with their staffs of OSCs. The Great Lakes region is expanded in Figure 4-2.

The most important aspect of a countermeasure is considered to be preparedness, and the OSC in each subregion is requested by the manual to develop effective emergency actions for his region. The following procedures are recommended:

1. Reporting Spills. An emergency telephone number should be established for the initiation of action in event of an emergency. This number should be widely publicized and active on a 24-hr-a-day basis. The polluter is primarily responsible for reporting a spill and initiating action.
2. Alerting Procedures. The OSC is responsible for assessing the magnitude of the spill and deciding the degree of involvement of local and provincial officials. It is suggested

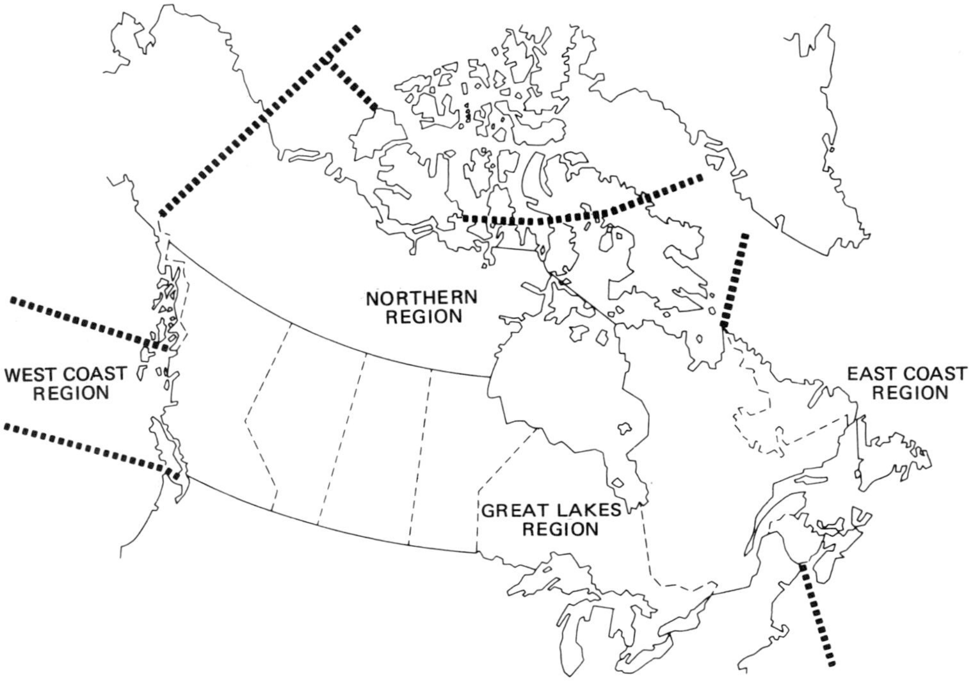

Figure 4-1 Regional boundaries, federal contingency plan. Source: Environment Canada, Interim Federal Contingency Plan for Oil and Toxic Material Spills, Ottawa, 1971, p. 10.

that municipal contingency organizations be established, especially by industry, and that these groups be prepared at all times.

3. Resources. Municipal contingency plans should involve organizations such as the Petroleum Association for the Conservation of the Canadian Environment (PACE) and the Canadian Chemical Producers Association (CCPA).
4. Federal Assistance. In event of a catastrophic spill, each OSC is instructed to request the resources of the Canadian Forces Operations Centre (CFOC).
5. Inventory of Local Hazards. Each OSC is requested to compile a list of potential pollutants that are moved through

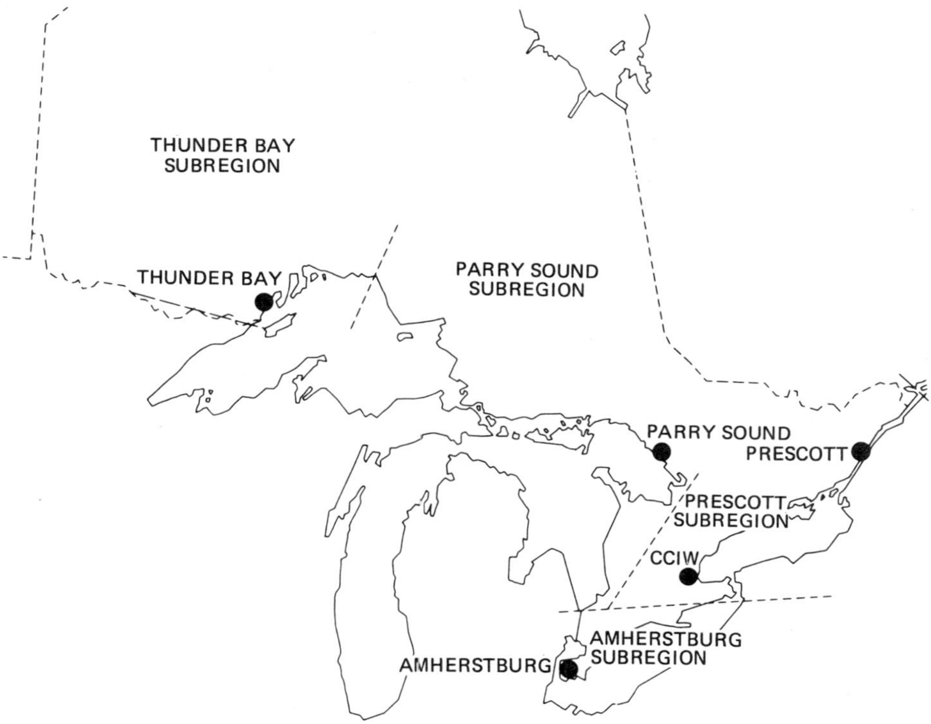

Figure 4-2. Great Lakes region, federal contingency plan. Source: Environment Canada, Interim Federal Contingency Plan for Oil and Toxic Material Spills, Ottawa, 1971.

or stored in the area. This should cover all types of transportation facilities including trucks, rail, marine, and air carriers of hazardous fluids and radioactive materials. Other areas for inventory involve chemical plants, warehouses, the frequency of equipment failures, etc.

6. Inventory of Available Facilities. Each OSC should know where the necessary equipment to combat spills can be obtained. Equipment of this kind should include bulldozers, earth moving machinery, oil booms, neutralizing chemicals, fill, straw, gravel, sawdust, disposal sites, etc.

7. Geographical Data. To plan for emergencies it is necessary

to have a thorough knowledge of wind and water currents in the area and downstream sites that might be adversely affected by spills of toxic substances. The location points of domestic water supplies, wildlife sanctuaries, recreational areas, etc., must be known.

8. <u>Preventative Measures</u>. OSCs are requested to initiate educational campaigns for local industry and those handling potential pollutants, and the general public should be instructed on potential hazards and how to react to emergencies.

The manual lists community facilities that can be used as countermeasures and sources of information. A section is devoted to countermeasures for oil spills, and TEAP (Transportation Emergency Assistance Plan) and CCPA (Canadian Chemical Producers' Association) are recommended as authorities regarding information on countermeasures for chemical spills.

4-1.2 The Ministry of Transport Plan for Marine Operations

The Ministry of Transport Plan [2] was written to guide federal officials in the design of effective emergency measures for marine spills. The manual was issued under the authority of the Deputy Ministers of Transport, Energy, Mines and Resources, National Defense, and National Health and Welfare, and the Ministry of Fisheries and Forestry. The prime objective of this plan is to guide in the removal of pollutants from the water environment and the cleanup of any damaging residue. Four procedures are defined:

1. Discovery and reporting
2. Containment
3. Removal
4. Cleanup

The waterways of all the provinces of Canada, the Arctic, and the United States are included. As a result, the provinces and the

International Joint Commission are involved in the plan. Spills in the Arctic and on the oceans are clearly federal problems, whereas a spill in a province might only require federal assistance.

The structure of the Ministry of Transport plan is generally the same as that of Environment Canada, where Canada is divided into regions with regional and on the scene coordinators. Where an emergency clearly involves one province, the plan relies on municipal and provincial officials to initiate action. Federal assistance is available through the Canadian Forces Operations Centre or the Ministry of Transport.

Federal response is coordinated through the Interim Interdepartmental Committee on Contingency Planning (IICCP) which is composed of representatives of the Federal Departments of Energy, Mines and Resources, Fisheries and Forestry, Indian Affairs and Northern Development, National Defense (including the Emergency Measures Organization), National Health and Welfare, and Transport. The committee is chaired by an official from the Ministry of Transport. The responsibilities of the committee are as follows:

1. To select OSC personnel and provide operational and technical support
2. To develop a mobile containment and cleanup team that can be at the scene of an emergency on short notice
3. To initiate research and development programs involving studies of the effects of pollutants on the environment and the design of emergency procedures
4. To initiate liason between provincial, federal, industrial, and other authorities

A Technical Working Group of the Interdepartmental Committee on Water has been established to assemble and develop knowledge on the best methods for combatting spills of hazardous materials. This group is chaired by a representative of the Department of Energy, Mines and Resources; its first assigned duty is to prepare a Contingency Plan Field Manual.

It is recommended that the polluter should pay for the cleanup of a spill. The cost of preparedness including personnel and material costs (stockpiling of chemicals, machinery, rentals, etc.) has to be borne by various government departments. Research costs fall under the Departments of Energy, Mines and Resources, Fisheries and Forestry, Indian Affairs, and Northern Development.

The recommended reporting procedure for an emergency is as follows:

1. The polluter or party that discovers the incident notifies a designated officer with a 24-hr contact capability. This initiates the actions of the OSC.
2. The OSC notifies a predetermined list of other officers, advisors, etc., who institute containment and cleanup procedures. This list is composed of individuals in the locality of the incident.
3. The OSC reports the incident to the Canadian Forces Operations Centre at Ottawa and this group in turn alerts designated officials.
4. The OSC and technical experts visit the scene of the incident for an assessment of the magnitude of the incident.
5. Ottawa is kept informed of the actions being taken through a public information officer from the OSC office. If additional resources are required, according to the judgment of the OSC, IICCP will comply by the fastest means possible.
6. A detailed operations log will be maintained.

4-1.3 The Province of Ontario Contingency Plan

The 1973 Province of Ontario Contingency Plan [3] outlines a framework for spill discovery and reporting, the coordination of manpower, equipment, etc., to control and clean up major spills. The plan is limited to spills affecting the surface or ground waters of the province and operates in cooperation with municipal and federal agencies.

A number of regional response teams have been established throughout the province to deal with emergencies. The main functions of these teams are to:

1. Establish effective communications between all parties related to the containment and the cleanup of spills.
2. Ensure that all necessary information is made available to the Response Team Coordinator during an incident.
3. Maintain an updated list of personnel, equipment, and other lists of resources.

Regional response teams have been formed in the Sarnia and Toronto areas, and are being developed at Thunder Bay, Sault Ste. Marie, Sudbury, Windsor, London, Hamilton, Toronto, Kingston, and other municipalities.

The Ontario Water Resources Act and the Environmental Protection Act make it mandatory that a spill of hazardous materials be reported as soon as possible. Failure to comply with this regulation is an indictable offense with a maximum fine of $5000. The ministry has the authority to organize and carry out a cleanup at the expense of the person or party responsible for the spill.

Spills are classified as being either minor, moderate, or major. For the cleanup of a minor spill, the party responsible can utilize his own resources and the public is not affected to any extent by the incident. Moderate spills can be dealt with effectively by the local contingency plan. Major spills require additional assistance. In the case of a major spill, the public and environment are subject to a hazard and the degree of concern by all parties is increased.

All spills are required to be reported to the appropriate regional staff of the Industrial Wastes Branch of the Ontario Ministry of the Environment. Minor spills can be reported during working hours; moderate spills, as soon as possible; and major spills must be reported immediately.

The plan gives detailed guidelines for cleanup of oil spills using chemicals.

4-1.4 Contingency Planning in the United States

Federal Planning

The Torrey Canyon and Ocean Eagle marine incidents and the Dunreith rail spill (see Section 3-3.1) initiated the formation of the 1968 National Multiagency Oil and Hazardous Materials Contingency Plan. It was the purpose of this plan to coordinate the efforts of various federal, state, and local agencies when dealing primarily with a large oil spill. The original plan has been amended several times to form the present national contingency plan.

On April 3, 1970, President Nixon signed the Water Quality Improvement Act; Sections 11 and 12 of this act deal with spills of hazardous substances and authorized the initiation of a National Contingency Plan. Two months later, the Council on Environmental Quality published a National Oil and Hazardous Materials Pollution Contingency Plan which was essentially the same as the 1968 plan. Agencies that became involved in contingency planning included the Departments of Transportation, Defense, and Interior plus advisory agencies including the Departments of Commerce, Treasury, and Health, Education and Welfare, and the Office of Emergency Preparedness. A national response team was developed to serve as a preparedness body and emergency-action force, and a staff of on-scene coordinators for various geographical areas was selected to oversee planning and handle incidents. Guidelines were defined for the initiation of action following a spill incident with regard to notifications, containment, countermeasures, cleanup and disposal techniques, restoration procedures, legal procedures for the recovery of damages, and the enforcement of environmental law.

On October 18, 1972, Amendment PL 92-500 of the Federal Water Pollution Control Act specified a series of penalty provisions for the discharge of hazardous substances and directed that the states should have access to the pollution revolving fund which financed the national plan. The Atomic Energy Commission became a participating agency. Since 1968, over $12 million was allotted to the development of a national contingency plan and the system has been invoked over 5000 times.

An active section of the National Contingency Plan is the Coast Guard National Strike Force (NSF) which has trained personnel and equipment to combat a spill. Teams of NSF personnel are available on east, west, and gulf coasts to assist on-scene coordinator groups. Four or more NSF personnel can reach a spill site in 2 hr and a full strength force of 19 men can be in action within 12 hr.

Vital components of NSF are the Chemical Hazards Response Information System (CHRIS) and the Air Deliverable Antipollution Transfer System (ADAPTS). Chemical, physical, and toxicological data on hundreds of hazardous substances are provided by CHRIS while ADAPTS defines the availability of pumps, hoses, winches, and other equipment required to contain and clean up a spill.

States Contingency Planning, An Overview

California and Wisconsin have well-established contingency plans while states such as Minnesota and Illinois have plans under development. The California plan involves federal, state, and local agencies and encourages participation by private industries. In the event of an emergency, a coordinator is selected from one of several state agencies, depending on the nature of the incident. Oil spills are handled by the Department of Fish and Game. The disposal of contaminated sediments is supervised by the Regional Water Quality Control Board.

Wisconsin's Contingency Plan for Spills of Oil and Other Hazardous Materials was first instituted in 1971. This plan is structured along the guidelines set by the federal plan, and funding for training programs and equipment is handled by the Department of Natural Resources.

A private firm (Ryckman, Edgerly, Tomlinson, and Associates) was contacted by the Minnesota Pollution Control Agency to design an inventory of state manpower and equipment resources for an emergency situation. This information will be incorporated into the state's contingency plan which is under development.

The Illinois plan involves the cooperation of various state and federal agencies in defining personnel and hardware for an emergency. The Illinois Environmental Protection Agency has two basic goals, one to implement a plan and the other to define preventative measures.

Railway Planning and Findings

The U.S. Railway System Management Association has developed a Hazardous Materials Manual to assist railway carriers in case of an emergency. This manual is continually updated through the cooperative efforts of the Bureau of Explosives, and the Federal Railroad Administration and Trade Associations, and ties into emergency data and action systems such as CHEMTREC, which was initiated by the Manufacturing Chemists Association.

In 1974, O'Driscoll of the Southern Railways System [4] provided the following information with regard to rail spills and transportation spills in general. Regarding the cause of rail spills, the following statistics were reported:

Cause of Spill	Percentage of All Spills
Leaking fill covers and loose gaskets	22
Leaking bottom outlets	16
Derailments	14
Ruptured or loose safety vents	9
Leaking drums and containers	8
Loose valves and plugs	5
Tank-shell corrosion	5
Overloads	5
Miscellaneous	16

The breakdown of total spills during 1973 was:

Type of Carrier	Total Spills
Air	48
Marine	12
Rail	412
Highway	5542
Pipeline	1637
Total	7651

From the above data it is apparent that most spills result during the highway transport of hazardous materials. Rail spills account for only about 5% of all spills. Of the 412 rail spills only 5% or about 21 spills were reported to have caused significant stream pollution.

4-1.5 Contingency Planning in Private Companies

Dow Chemical Company

Dow is one of the largest chemical producers on the North American continent and millions of kilos of many different chemicals are transported daily throughout Canada and the United States.

Considering the hazardous nature of many of these products, Dow has developed a plan to deal with either transportation or in-plant emergencies. The communications structure of this plan is shown in Figure 4-3.

The key figures in the Dow plan are the Knowledgeable Dow Contact (KDC) and the Emergency Coordinator (EC). On notification of an emergency, the KDC decides what action must be taken and the EC assists in this decision-making process. Emergency telephones are located throughout Dow plants in every major division and these phones are manned 24 hr per day.

It is the responsibility of KDC personnel to prepare for emergencies and assemble data on hazardous substances. Individual cards are maintained for individual chemicals listing chemical, physical,

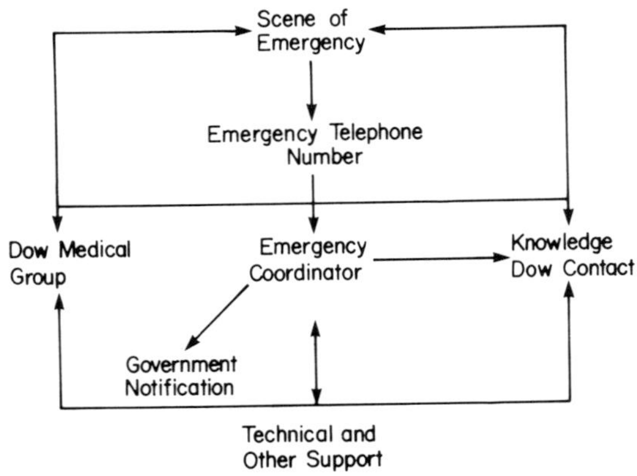

Figure 4-3 Example communication system. Dow Chemical communications system for emergencies involving products. Source: J. M. Theis et al., An Industry Distribution Emergency Response System, Proc. 1974 National Conf. Control of Hazardous Material Spills, San Francisco, August 25-28, 1974, p. 46.

and toxicity data, and actions that must be taken in case of emergencies to protect humans and their environment. Various groups such as medical and technical resources assist during spill crises. Dow personnel have traveled as far as South America, Africa, and Australia to assist in emergencies.

Rohm and Haas Kentucky, Inc.

This company has developed a system to deal with in-plant spills (Figure 4-4). The system requires that the plant have both chemical and storm sewers. It must first be decided whether the spill can be contained by diking and if not, whether the overflow can be diverted to a chemical sewer. The plant fire department assists in fire protection and preventing personnel injuries. If materials are directed to the chemical sewer system, the waste-water treatment department must be notified to initiate special storage and treatment processes. The State Environmental Protection Agency, Coast Guard, and/or Stream Pollution Authority must be notified if a spill is of sufficient magnitude to cause the pollution of a waterway. State and corporate resources then combine to handle the crisis.

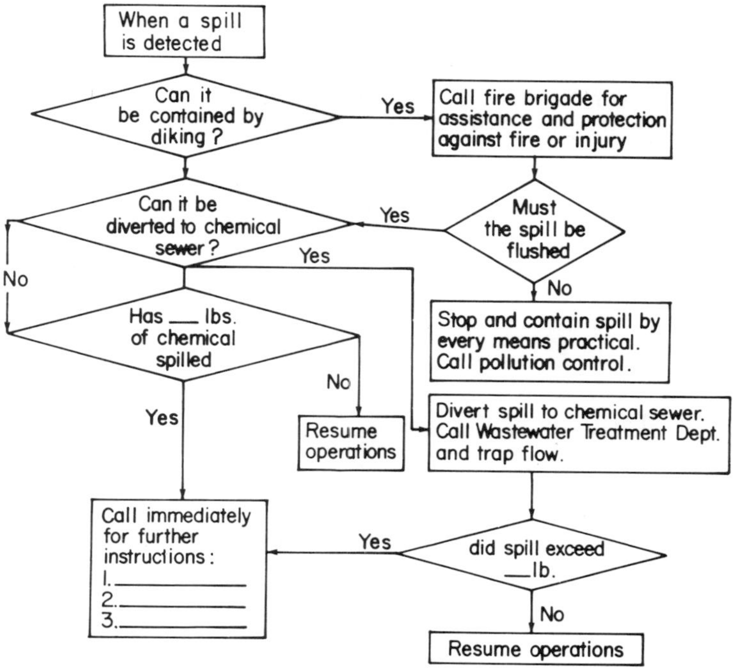

Figure 4-4 Rohm and Haas Kentucky, Inc., in-plant emergency spill control system. Source: R. A. Jensen, Spill Control Within a Chemical Plant, Proc. 1974 National Conf. Control of Hazardous Material Spills, San Francisco, August 25-28, 1974, p. 65.

4-1.6 Action Steps for a Warehouse Fire Involving Highly Hazardous Materials [5]

1. Should a fire occur, company personnel must notify:
   a. The fire department. The fire department should be advised of hazards involving:
      (1) toxic smoke and vapors
      (2) contaminated run-off water
      (3) the general toxicity of the materials involved
      (4) the flammability and explosive nature of the substances involved

b. The police department. The police department should be assisted by:
   (1) blocking off the area
   (2) advice for the necessity of evacuating neighboring residential and business personnel
   (3) advising of dangers concerning contaminated run-off water
   (4) suggesting possible support agencies
c. Transport agencies to remove transport trucks, railcars, etc., from the emergency area
d. Government pollution control agencies. Advise these agencies on the type and quantity of hazardous materials involved and on decontamination and disposal plans.
e. The municipal waste treatment facility. A surge of contaminated run-off water from the fire might upset the biodegradation process at the waste-treatment plant. If notified immediately, municipal personnel can take preventative measures such as closing off a sewer branch or directing influent to a holding pond.

2. It must be assured that toxic substances will not contaminate the municipal water supply. Warehouse water lines must be shut off to prevent back-up.
3. Steps must be taken to divert or dam up run-off water if there exists danger of human exposure to toxic materials, or sewer or stream contamination.
4. Toxic smoke or vapors must not contaminate food products in adjacent areas.
5. Call for assistance from various municipal and state or provincial emergency groups.
6. Provide for the protection of personnel involved in cleanup operations by supplying access to showers, protective clothing, medical facilities, rest cots, and information for hospital treatments.
7. Make preparation for the disposal of contaminated debris:

   a. arrange for a burial site
   b. prepare a hardware inventory (trucks, backhoes, chemicals, etc.)
   c. define the appropriate government agency for the supervision of burial procedures.

8. The emergency site must be decontaminated, unless otherwise specified (1% bleach solution is a very effective scrubbing agent).
9. Arrange for a watchman to guard the site during cleanup operations.

4-1.7 Municipal Emergency Planning: City of Sarnia, Ontario

A high concentration of chemical industries exists at the south of the city of Sarnia extending beyond the city limits, southward along the St. Clair River. A number of past emergencies have caused this municipality to formulate an emergency plan in the form of By-law No. 6889 (1973). This by-law defines the responsibilities of various city, public and private, officials when an emergency occurs. The goals of the by-law are to safeguard the public, maintain order, coordinate general operations involving rescue, evacuation, medical and social services, and to provide a vehicle of communication between emergency teams. All emergency operations are directed from the city police department by an Emergency Operations Control Group composed of:

The mayor
City manager and deputy city manager
Police chief
Commissioner of works
Commissioner of parks and recreation
Director of community family services
Building inspector

Purchasing agent

Medical officer of health

Managers of electrical, gas, and telephone services

The emergency measures coordinator

In the case of industrial emergencies or emergencies involving hazardous materials, the Chemical Valley Emergency Control System (CVECS) is activated. This group is composed of representatives from the major industries in the area who maintain a state of readiness to initiate a complex operation involving highly trained personnel and a pool of equipment in times of emergency. Monthly meetings are held by this group to assess responses to previous emergencies and to test responses to simulated emergencies.

CVECS is closely involved with CCPA who have developed TEAP which is designed to provide technical advice and assistance to police, fire, and civil authorities regarding highway, rail, and marine accidents involving chemical substances.

A municipal by-law (No. 6889) designates city and provincial police as responsible for incidents involving explosives or explosive devices.

Oil spills and spills of hazardous materials come under the jurisdiction of the Lambton County Regional Operation Team (ROT) which is a part of the Ontario Ministry of the Environment's Contingency Plan for Spills of Oil and Other Hazardous Materials.

Incidents involving radioactive materials are the responsibility of the Radiation Protection Service of the Provincial Ministry of Health. The Emergency Measures Coordinator assists in this area.

The by-law mandates the listing of phone numbers of all personnel and agencies which may potentially become involved in an emergency. Lines of communication and the authority of various personnel are clearly defined. In summary, this by-law is the official definition of a successful and working emergency planning system.

4-2 EMERGENCY DATA AND ACTION SYSTEMS

Many emergency data and action systems [6] have been developed in the United States and Canada over the past decade. A good deal of overlap and duplicated effort is evident in these systems. It appears that a computerized system such as OHM-TADS, which is discussed below, will eventually dominate other data systems.

TEAP [7]

The Transportation Emergency Assistance Program (TEAP) was developed by the Canadian Chemical Producers' Association and introduced on January 15, 1971. A team of chemists and engineers and personnel familiar with transportational equipment were assembled across Canada to respond to emergencies on a 24-hr basis. Industries providing this expertise are listed in Table 4-1.

Table 4-1

The Canadian Chemical Producers' Association Transportation Emergency Assistance Plan Regional Control Centers

Company	Location	Geographic Area
Hooker Chemicals Division	Vancouver	British Columbia
Celanese Canada Ltd.	Edmonton	Prairies
Canadian Industries Ltd.	Copper Cliff	Northern Ontario
Dow Chemical of Canada Ltd.	Sarnia	Central Ontario
Cyanamid of Canada Ltd.	Niagara Falls	Eastern Ontario
DuPont of Canada Ltd.	Maitland (Ont.)	Quebec and south of St. Lawrence
Allied Chemical Canada Ltd.	Valleyfield	Quebec and north of St. Lawrence
Gulf Oil Canada Ltd.	Shawinigan	Quebec

Source: E. N. Banks, The Canadian Chemical Producers' Association Transportation Emergency Assistance Plan, Proc. 1974 National Conf. Control of Hazardous Material Spills, San Francisco, August 25-28, 1974, p. 73.

When a call is received, pertinent information is recorded after which a technical adviser is assigned to handle the emergency. The adviser coordinates information provided by the manufacturer of the chemical substance involved in the spill and the actions of men and equipment which are on the scene. The adviser has access to a pre-developed communications network, a TEAP manual listing chemicals, their manufacturers, and various assistance agencies and a series of standard references giving emergency information. A total of 134 incidents have been handled by TEAP through July of 1974.

The TEAP plan has become part of the emergency plans of the Federal Contingency Plan, the Ontario Department of the Environment Contingency Plan, the Canadian National Railway and the Canadian Pacific Railways Plans, and the Quebec Fire Code. TEAP is also allied with the CHEMTREC and CHLOREP Plans, the Canadian Agricultural Chemicals Association, and the Compressed Gas Association.

CHEMTREC

The Chemical Transportation Emergency Center system became operational in September of 1971. CHEMTREC provides 24-hr telephone service to on-scene personnel supervising emergencies. The CHEMTREC dispatcher may coordinate or maintain contact with other emergency systems such as CHLOREP, TERP, etc. (See Appendix B, Table B-1.)

OHM-TADS

The Oil and Hazardous Materials Technical Assistance Data System was put into practice in January of 1972. This system was developed by the United States Environmental Protection Agency and is a computer-based information depository which provides information on how to treat spills involving more than 850 substances including oils, dispersants, and hazardous materials. OHM-TADS is, perhaps, more pertinent to environmental protection than either TEAP or CHEMTREC which tend to emphasize first aid. In the OHM-TADS concept, feedback from field personnel is immediately computerized to provide decision-making guidance. The system is continually updated to include accident research findings and to include information on new products and new methodology.

WATDOC

The Water Resources Document Reference Guide Centre (WATDOC) of Environment Canada is a cooperative computerized data information project initiated within the Environmental Management Service by the Inland Waters Directorate (Burlington, Ontario). Federal and Provincial agencies together maintain an updated coverage of scientific and technical literature dealing with water resources. This information is based upon abstracts, bibliographic citations, economic studies, sociological research, management reports, statistical tables and news clippings, etc. Information is shared between Canada and the United States. At the present time over 25 governmental, industrial, and university groups share in WATDOC Data Resource Centre. (Information from Environment Canada Bulletin on WATDOC, April 1974.)

REFERENCES

1. Environment Canada, Interim Federal Contingency Plan for Oil and Toxic Material Spills, Field Manual, Ottawa, 1971.

2. Ministry of Transport, Interim Federal Contingency Plan for Combatting Oil and Toxic Material Spills, Marine Operations, Ottawa, 1972.

3. P. G. Selling, The Province of Ontario Contingency Plan for Spills of Oil and Other Hazardous Materials, Status Report, Ontario Ministry of the Environment, Toronto, 1973.

4. J. J. O'Driscoll, Spill Prevention and Control in the Railroad Industry, Proc. 1974 National Conf. Control of Hazardous Material Spills, San Francisco, August 25-28, 1974, pp. 140-142.

5. Environmental Protection Agency, The Pollution Potential in Pesticide Manufacture, Technical Studies Report TS-00-72-04, U.S. Govt. Print. Off., Washington, 1972, pp. 239-242.

6. G. J. Moein, C. H. Thompson and G. W. Dawson, Emergency Data for Hazardous Substances, 28th Annual Purdue Industrial Waste Conf., School of Civil Eng., Purdue Univ., Lafayette, Ind., May 1-3, 1973.

7. E. N. Banks, The Canadian Chemical Producers' Association Transportation Emergency Assistance Plan, Proc. 1974 National Conf. Control of Hazardous Material Spills, San Francisco, August 25-28, 1974, pp. 72-74.

Chapter 5

UNITED STATES AND CANADIAN ENVIRONMENTAL LEGISLATION AND PROCESSES

5-1 UNITED STATES LEGISLATION

The U.S. federal legislation dealing with hazardous substances and pesticides and how they relate to use, sale, transportation, and application include the following:

1. Water Quality Improvement Act, P.L. 91-224, 84 Stat. 91 (1970)
2. Environment Quality Improvement Act, P.L. 91-224, 84 Stat. 114 (1970)
3. Clean Water Restoration Act, P.L. 89-733, 80 Stat. 1246 (1966)
4. Water Quality Act, P.L. 89-234, 79 Stat. 903 (1965)
5. Federal Water Pollution Control Act, P.L. 87-88, 75 Stat. 204 (1961)
6. Water Pollution Control Act, P.L. 84-660, 70 Stat. 498 (1956)
7. Water Pollution Control Act, P.L. 82-579, 66 Stat. 755 (1952)
8. Water Pollution Control Act, P.L. 80-845, 62 Stat. 1155 (1948)
9. Fish and Wildlife Coordination Act, P.L. 85-624, 16 USC (1958)

10. Federal Insecticide, Fungicide and Rodenticide Set, 7 USC 135, 61 Stat. 163 (1919)

Important executive orders dealing with hazardous substances and pesticides and their effects include:

1. Executive order 11507--Prevention, control and abatement of air and water pollution at federal facilities, 4 Feb. 1970.
2. Executive order 11288--Prevention, control and abatement of water pollution by federal activities, 2 July 1966.

By executive order 11574 of December 23, 1970 a new federal program to control water pollution from industrial sources through use of the permit authority in the Defense Act of 1899 was established. This was intended to give a more efficient means for abating discharges from industrial plants such as endrin discharges that caused the 1963-64 Mississippi fish kills.* The Refuse Act outlaws discharges and deposits other than municipal sewage into all navigable waters whether they are intentional or unintentional.

A summary of state pesticide use and application laws may be found in the report "Pesticide Study Series 11," which summarizes the laws and institutional mechanisms controlling the release of pesticides into the environment [5].

The finest single document covering legal cases on environmental law is that by Bockrath [1]. The Federal Water Pollution Control Act (FWPCA) as passed in 1948 authorized the surgeon general of the U.S. Public Health Service to make studies of water-borne pollution sources, setting standards of water quality, and made it clear that pollution control responsibility rested with the states. This responsibility changed and was transferred to the administration of the Environmental Production Agency with the Clean Water Restoration Act of 1966. The Water Quality Improvement Act of 1970 (WQIA) repealed the Oil Pollution Act of 1924 and added several sections, e.g., pollution by hazardous substances, mine water pollution, sewage

*U.S. Public Health Service, Transcript of the Conference on the Pollution of Interstate Waters (Lower Mississippi River) at New Orleans, March 5-6, 1964.

discharge from vessels, and pollution control in Alaska and the Great Lakes. The present stage of evolution of the FWPCA is embodied in amended form in 1972 (33 USC 1251 et seq.). It regulates point sources of pollutant discharges, controls spills of oil and hazardous substances, and provides financial assistance for municipal waste treatment [10]. The "Ocean Dumping Act," entitled "The Marine Protection, Research and Sanctuaries Act of 1972," (USC 1401 et seq.) established a permit system for disposal of wastes at sea.

Environmental impact requirements are covered in the United States under the National Environmental Policy Act of 1969 (NEPA).* This piece of legislation is the most far-reaching of any that is in existence today. The stated purpose was to declare a broad mandate and to establish a comprehensive policy for the environment:

> to declare a national policy which will encourage productive and enjoyable harmony between man and his environment; to promote efforts which will prevent or eliminate damage to the environment and biosphere and stimulate the health and welfare of man; to enrich the understanding of the ecological systems and natural resources important to the nation; and to establish a council on Environment Quality.

To implement and promulgate the act, the Council on Environmental Quality (CEQ) prepared guidelines for federal agencies. This activity aided those active in preparing environmental impact statements, but of even more importance, the CEQ initiated a new publication, the 102 monitor, which reviewed all court decisions involving NEPA. Many of these decisions involved "more hazardous wastes" [17]. Brown [2] examined the problems of applying NEPA to joint federal and nonfederal projects.

In 1959 the 1947 pesticide act was amended to include nematocides, plant regulators, defoliants, and desiccants, and in 1964 another amendment eliminated the controversial "registration under protest" provision of the 1947 act and authorized the secretary of

---

*42 United States Code 4321 et seq. (Supp. 1970) Public Law 99-190, Jan. 1, 1970.

agriculture to require labels to bear their federal registration number.

The hazardous materials transportation control legislation placed the responsibility of regulation and enforcing the law regulating the interstate shipment of hazardous material in the Department of Transportation (DOT) by the DOT act of 1966 (see Sections b, c, and e, 49 USC 1655). Under this act the powers formerly found in three separate agencies, i.e., The Interstate Commerce Commission, The Federal Aviation Agency, and the Coast Guard (18 USC 831-835; 46 USC 170; and 49 USC 1421-1430, 1471, 1472h) were transferred to the DOT.

The hazardous material regulation board have classified class B poisons as those which are so toxic to man as to present a human health hazard during transportation. They must carry a special label on the outside of the package. Neither class A nor B poisons may be transported or stored in vehicles with foodstuffs or animal feeds (49 CFR 173, 402 et seq.; 49 CFR et seq.; 49 CFR 177-841e).

The legal framework for controlling the release of pesticides into the environment consists of federal laws, executive orders, state laws, and institutional mechanisms, international law and interintraagency organizations devised by federal agencies for coordinating pest control.

The federal laws provide indirect control of varying degrees of effectiveness on both use and release of pesticides, e.g.: (1) registration of pesticides products for interstate distribution, (2) regulation of amounts of specific pesticides residues tolerated in raw agricultural products and processed foods, (3) establishment of state-federal water quality standards limiting toxic substances in interstate waters, (4) provisions limiting toxic substances in interstate waters and provisions for surveillance of the environmental impact of all federally supported pest control programs, (5) provisions for research and monitoring the effects of pesticides on man and for training public officials and provisions for investigating effects of pesticides on fish and wildlife and in dissemination of results.

This all sounds good. However, with the exception of the Federal Aviation Administration, which regulates crop-dusting activities, the federal laws are not really effective in regulating the use of pesticides. An older law, the refuse act of 1899, does provide comprehensive regulation of industrial water pollution. Also, executive order 11507 provides for the control of water and air pollution at federal facilities. However, it still must be admitted that U.S. federal laws are ineffective in regulating the release of waste pesticides into the environment.

Forty-nine states have statutes for registration and labeling of pesticides for the lawful sale and distribution of these materials. At least 20 states impose use restrictions, and in 19 states licenses or permits are required for pesticide dealers. Further, 31 states have statutes requiring the licensing of commercial or custom pesticide applicators (Pesticide Study Series 11, 1971); almost all state regulation must be related back to the Federal Insecticide, Fungicide and Rodenticide Act (FIFRA), 61 Stat. 163, as amended, 7 USC 135-135K, which was originally passed in 1947 to regulate the marketing of "economic poisons" and "devices" and was amended again in 1959, 1961, and 1964. The term economic poison has the same meaning as pesticide. Devices are mechanisms such as ant traps, sold together with pesticides for the purpose of application, or simply electronic bug killers that are designed to kill pests. Under FIFRA, all pesticides and devices must be registered if they are to be legally shipped in interstate commerce.

Section 408 of the Federal Food, Drug and Cosmetic Act (FDCA), the so-called Miller Amendment, was passed in 1952 and it authorizes the administrator of the EPA to establish residue tolerances or exemptions from tolerance of pesticides. Some changes in interpretation of Section 408 have taken place in Section 409 of the FCDA and addition of the Delaney clause of the food additives amendment, which states that no "food additives" capable of causing cancer in animals or man may be added to food (21 USC 348; 21 USC 348c). At least one exception to the Delaney clause may be found since it <u>may not</u> apply to the use of DDT on raw agricultural products (428 F2d 1083, D.C. Cir. 1970).

There is very little information relative to legal aspects of pollution in the Proceedings of the 1974 National Conference on

Control of Hazardous Material Spills. Morgenstern [9] has carefully examined the relationships between federal and state laws for controlling and preventing pollution while the state environmental protection legislation and the commerce clauses are reviewed in Harvard Law Review 87: 1762 (1974) and in a review by Yost [18].

The Massachusetts law relating to hazardous substances is reviewed in a "Pesticide Study Series 9" (1972).

The International Joint Commission (IJC) has been established to improve the management of the Great Lakes by the United States and Canada. The power vesting in the IJC and how management is accomplished may be found in the report by Zile [19].

5-2 CANADIAN LEGISLATION

Various government agencies have been established to enforce environmental statutes. These agencies fall under the jurisdiction of federal, provincial, and municipal levels of government. They are further divided into branches of government, including the legislative branch that passes the statutes, the judiciary which interprets the law, and the executive which administers the law. The following discussion is primarily concerned with the statutes that serve to indict those who spill highly hazardous substances and the processes that exist for the recovery of damages by persons affected by spills.

The Canadian federal government has the legal authority to pass laws concerning matters that are considered national in scope under the British North America Act. Such areas involve shipping, navigation, fisheries, etc. To improve the effectiveness of federal authority, Environment Canada was established in 1972 with six prime goals:

1. Carry on established resource programs and services
2. Clean up and control pollution
3. Assess and control the environmental impact of major development
4. Initiate long-term environmental programs
5. Promote and support international environmental initiatives and
6. Develop an environmental information and education program.

One of the most useful accomplishments of Environment Canada was the initiation of contingency planning that was discussed in Section 4-1.1.

Probably the most useful federal statutes with regard to protection against spills of highly hazardous materials include: the Canada Shipping Act, the Fisheries Act, the Criminal Code, and the Environmental Contaminants Act. Quotes from these statutes and various details are listed in Appendix C.

The Canada Shipping Act is designed to control pollution caused by ships and vessels operating in Canadian waters. The maximum fine for the negligent or willful discharge of "oily materials," pesticides, etc., into Canadian waters is $100,000. Prosecution can include the owner of the ship as well as the ship's master. Ship owners are required to carry limited liability up to a maximum of $14 million to cover the cost of pollution cleanup property damage, which includes recompensation of the loss of income for fishermen, and cargo, as well as wreck removals. A Maritime Pollution Claims Fund exists for unsatisfied judgments with respect to pollution damage exceeding liability limits. Failure to provide evidence of financial responsibility or the failure to report a pollution incident can result in a maximum fine of $100,000.

One of the most important federal statutes for combatting water pollution is the Fisheries Act. It is primarily concerned with pollutants that are injurious to fish. Polluters are liable to fines up to $5000 for each offense committed under the sections of the act. Unique features of the act include:

1. The right of the courts to order the stopping of procedures that cause pollution along with a fine and imprisonment or very high fines, if the polluter does not comply with the orders of the court.
2. The right of a citizen, who initiated the action that led to a conviction, to share one half the fine plus one half of the proceeds recovered from the sale of any equipment seized and sold by the government in connection with the violation.

3. A two-year maximum period for laying a charge against the polluter.

Section 387 of the Criminal Code of Canada defines the conditions under which a polluter can be charged with mischief causing property damage or danger to life. The period for imprisonment with respect to property damage is up to five years, while a penalty of life imprisonment can be imposed for mischief causing danger to life.

The federal government has recently drafted an Environmental Contaminants Act which is expected to become law in the near future. This act will empower the Federal Cabinet to "regulate, prohibit or restrict the importation, manufacture, distribution, sale, use or release of any substance or class of substances which pose or are believed to pose an unacceptable hazard to the environment or to human health and well-being" [7].

Other federal statutes under which perpetrators of spills can be charged are listed below. The National Harbours Board Act covers spills from ships in harbor. Minor fines and imprisonment up to six months can be levied under the Migratory Bird Regulations Act. Failure to conform with the regulations of the Railway Act could result in a relatively small fine up to a maximum of $500. Violations of regulations under the Aeronautics Act could result in a fine up to $5000. The following are Canadian federal statutes under which perpetrators of spills can be charged:

Aeronautics Act
Atomic Energy Control Act
Canada Shipping Act
Criminal Code
Fisheries Act
Government Harbours and Piers Act
Harbour Commissions Act
Hazardous Products Act
Migratory Birds Convention Act
National Parks Act
Navigable Waters Protection Act
Nuclear Liability Act

The International Joint Commission specifies that adequate protection must be provided for the removal or reduction of toxic substances in international waterways. It is specified that the phenol concentration of water should not exceed 5 ppb, iron 0.3 ppm, and pH values shall range between 6.7 and 8.5. Dissolved oxygen values should not fall below 3 mg/liter. IJC legislation is basically a statement of international agreements rather than a vehicle for prosecution.

A review of provincial statutes reveals that the provinces of Ontario and Alberta are among the foremost of the provinces with respect to environmental legislation. Two of Ontario's strongest acts are the Ontario Water Resources Act (OWRA) and the Environmental Protection Act (EPA).

The basic scheme of the OWRA of 1972 centers on certificates of approval for municipal and industrial uses of water and the discharge of pollutants into the water. In theory, the limits that have been approved by the ministry will maintain safe levels of contaminants in the environment. Section 32-(1) of the OWRA (see Appendix C-9) states that every person who discharges into water material that adversely affects the quality of the water is liable on first conviction to a fine of $5000 or less, and on each subsequent conviction to a fine of not more than $10,000 and/or imprisonment up to one year. It is specified in Section 32-(4) of the OWRA that the failure to immediately notify the ministry of a spill can result in a fine of up to $5000. Unfortunately, no quantitative limits for pollutants have been set in this act for a meaningful definition of adverse contamination. Therefore, prosecution may be difficult. Several important judicial interpretations of the act, which have established legal precedent, are as follows:

1. It is not necessary to establish that a material "did in fact impair, but rather that it had the ability to do so ...." [13].
2. Proof of "mens rea" is not required for conviction, and ignorance is no excuse for pollution [12].

3. A person may be guilty of an offense if his agents or employees had the authority to prevent pollution and did not do so [13].

The Canadian EPA (1971) gave the Ontario Ministry of the Environment authority to investigate problems dealing with pollution and authority to take action. One such action is the demand that anyone transporting hazardous materials maintain the essential equipment required to meet emergencies. Should a spill occur, the polluter can be billed for the cleanup. A procedure is provided for any citizen to request an investigation of matters dealing with pollution and to seek reimbursement for damages associated with pollution. A maximum fine of $5000 is specified for a first offense and $10,000 for subsequent offenses.

The province of Alberta's strongest legislation is the Clean Water Act (Appendix C, No. 14). Persons causing a spill or discharge of contaminants are required to report the incident within 24 hr of its discovery and issue a comprehensive report within 72 hr. Failure to meet these requirements can result in a fine of $1000 per day for each day that the contravention continues. The Clean Air Act specifies parallel requirements for air emissions (Appendix C, No. 17).

Alberta requires under regulations OC 214/73 and OC 217/73 that designs for new plants include emergency procedures for the prevention of accidental releases of environmental contaminants (Appendix C, Nos. 15 and 16).

The Public Health Act of Alberta specifies that no person shall discharge chemical substances into sewers or cause hazardous substances to enter public waters. A list of criteria is presented for various toxic substances. Penalties are small, being on the order of from $50 to $200. Penalties under the Litter Act are also small, and short terms of imprisonment are included. Further environmental protection is offered under the Municipal Government Act of Alberta (1968), but penalties are left to the judgment of the courts.

A large amount of authority is provided for the Director of Pollution for the province of British Columbia in the Pollution Control Act of 1967. The director has been given necessary power to define what constitutes a pollution problem and to issue orders to repair, alter, or improve an operation. The Director of Pollution Control can enter the premises of any industry at any time and, if necessary, close down the operation. Those convicted of an offense under this act are liable to a fine of up to $1000 and/or imprisonment up to three months. For each day the offense continues, a fine up to $500 can be levied. Further legislation is provided by the Health Act.

Selected statutes for the provinces of Manitoba, Quebec, New Brunswick, and Nova Scotia are presented in Appendix C, Nos. 20 to 27. In Manitoba, anyone causing a spill could be prosecuted under the Clean Environment Act, the Public Health Act, or the Municipal Act. Penalties are rather minor. Environmental legislation in the provinces of Quebec and New Brunswick is weak, and prosecutions are possible under the Water Board Act and the Cities and Towns Act in Quebec and the Water Act in New Brunswick. Nova Scotia is protected by the Public Health Act and the Environmental Pollution Control Act

A riparian landowner can sue for damages resulting from a pollution incident under common law rights. Private nuisance can be established by the owner of a property if the pollutants cause unnecessary interference with the use of the land and economic losses or health damage. Persons or companies are liable to damages if negligence can be proven; however, this is generally more difficult than establishing nuisance. The courts have established precedent that any person permitting harmful materials to escape from his land does so at his own peril and will be held responsible for any resultant damages [15].

In summary, perhaps the strongest Canadian federal statutes serving as a deterrent against the spilling of hazardous materials are found in the Fisheries Act and the Canada Shipping Act. Two statutes in the Province of Ontario Legislation, the Ontario Water Resources Act and the Environmental Protection Act, as well as the Clean Water Act of Alberta, represent the strongest provincial statutes. Many other federal and provincial statutes have been discussed, but the penalties defined in these acts are relatively minor. Most statutes are weak in the sense that they have failed to define quantitative limits for pollutants.

REFERENCES

1. J. Bockrath, Environmental Law for Engineers, Scientists and Managers, McGraw-Hill, New York, 1977.
2. E. Brown, Applying NEPA to Joint Federal and Non-Federal Projects, Environmental Aff., 4: 135 (1975).
3. Control of Hazardous Material Spills, Proceedings of the 1974 National Conference on Control of Hazardous Material Spills, August 25-28, 1974, San Francisco, California. Sponsored by A.I.Ch.E. and the U.S. EPA.
4. Environmental Education Act, P.L. 93-278, 88 Stat. 121 (1974).
5. EPA 1971 Pesticide Study Series 11, Laws and Institutional Mechanisms Controlling the Release of Pesticides into the Environment, pp. 1-140. EPA Form 1510-2 (6-71).
6. EPA, Pesticide Study Series, The Movement and Impact of Pesticides Used for Vector Control on the Aquatic Environment in the Northeastern United States, pp. 1-202, Contract No. 68-01-0129.
7. D. Estrin and J. Swaigen, Environment on Trial, The Canadian Environmental Law Research Foundation, Toronto, 1974.
8. Federal Water Pollution Control Act as amended. 33 USC 1151 et seq., Oil Spill Provisions at 33 USC 1321.
9. A. Morgenstern, Relationship Between Federal and State Laws to Control and Prevent Pollution, Environmental Law, 1: 238 (1970).
10. J. W. Raisch, Enforcement Under the Federal Water Pollution Control Act Amendments of 1972, Land & Water L. Rev., 9: 369 (1974).
11. Refuse Act, 33 USC 407 (more completely identified as the Rivers and Harbors Appropriation Act of 1899), 33 USC 401 et seq.
12. R. v Industrial Tankers Ltd. (1968) 2 or 142 at 150. (Quoted with approval in R. v Sheridan, Ref. 3.)
13. R. v Matspeck Construction Co. Ltd., Ont. Prov. Ct. (1955-650 6 Crim. L.Q. 455 at 460). (Quoted with approval in R. v Sheridan, Ref. 3.)
14. R. v Sheridan (1973) 2 or 192 at 204 (Dist. Ct.).
15. Rylands v Fletcher (1866) LR 1 Ex 265, Affirmed (1868) LR 3 HL.
16. Toxic Substances Control Act of 1973, 93 Congress, Calendar No. 240, S. 426, Report 93-254.
17. V. J. Yannacone, Natural Environmental Policy Act of 1969, Environmental Law, 1: 8 (1970).
18. N. C. Yost, NEPA's Progeny: State Environmental Policy Acts, Environmental Law Rep., 3: 50090.
19. Z. L. Zile, Binatural Land Resource Management for the Great Lakes Area: Powers of the International Joint Commission. Working Document No. 1: Great Lakes Management Series, 1974.

Chapter 6

CONTAINMENT AND TREATMENT TECHNIQUES FOR SPILLS

No two emergencies are the same. A spill can occur at a plant during a processing or loading operation or in the field during the transit of materials. Facilities to cope with emergencies are often more readily available in plants than in the field. During field spills, a good deal of damage might be caused to the environment before containment and treatment procedures begin. During an emergency involving a spill, one must assess:

1. The volume of hazardous substance released to the environment and the rate of leakage still in progress,
2. What danger exists to personnel in the area,
3. The nature of the damage and what repairs might be attempted,
4. Whether transfer to an alternative container is advisable,
5. Whether some form of dike should be constructed,
6. The nature of the spill area and how the released substance has spread over the area,
7. Whether the spilled substance can reach or has reached a watercourse or sewer,
8. The danger of explosion and fire,
9. What effect rain and wind will have on the spill, and
10. What equipment and supplies are necessary to confine the substance and carry out the necessary treatments.

Hazardous substances that enter a waterway either directly or by way of ditches or a storm sewer can produce a number of undesirable effects including fish kills, upsets in the natural flora and fauna of the aquatic environment, rendering the water unsuitable for consumption by man and animals, damage to recreational facilities, and the creation of a fire or air-pollution hazard. Spilled material that flows into a sanitary sewer can produce serious consequences with regard to the functioning of aerobic and anaerobic digestion processes at a sewage treatment plant. The safety of the human population of the area and containment of a spilled substance are of the first priority.

Hazardous substances may be in the form of a solid, liquid, or highly volatile material or gas. The containment of a solid on the ground may involve only scooping up the substance and transporting it to a disposal area. On the other hand, a cleanup may be complicated by explosions, wind, rain, or contamination with other compounds. Fire in the presence of nitrated organic compounds, ammonium nitrate, or chlorates may result in explosions with the release of toxic vapors.

It is essentially impossible to contain hazardous vapors in the atmosphere. The only logical approach is to stop the leak, and, if possible, insulate the volatile substance with foam. Care must be taken that inhabitants in the area and workers are not exposed to toxic vapors or endangered by a potential explosion. When vapors are flammable, open flames and arcs must be banned from the area.

Hazardous substances may contaminate groundwater following the seepage of spilled material into the soil. The hazards of groundwater contamination have been discussed by Metry [1], who emphasizes the importance of modeling the migration of hazardous substances in groundwater aquifers. Subsurface waters may become contaminated as a result of hazardous wastewater spills on land, from the seepage of a hazardous substance into the soil, through deep-well injection, or from leachates released by hazardous solid wastes.

The simultaneous molecular diffusion, convective dispersion, and chemical reaction model of aquifer dispersion processes are given by:

$$\frac{\partial c}{\partial t} + u\frac{\partial c}{\partial x} + v\frac{\partial c}{\partial y} + w\frac{\partial c}{\partial z} = \frac{\partial}{\partial x}\left(\sigma_x \frac{\partial c}{\partial x}\right) + \frac{\partial}{\partial y}\left(\sigma_y \frac{\partial c}{\partial y}\right) + \frac{\partial}{\partial z}\left(\sigma_z \frac{\partial c}{\partial z}\right) - f(c)$$

Hydrogeologic parameters that govern the diffusion of hazardous substances in aquifers include:

1. The initial concentration of the substance (c),
2. The time (t) for the substance to build up in the aquifer, achieve a steady state of concentration, and effect recovery as the hazardous substance is removed from the area.
3. The effective diffusion coefficients σ_x, σ_y, σ_z. The coefficient σ_x has been found to be directly proportional to the first power of groundwater flow, whereas the lateral coefficients have the order of magnitude of molecular diffusion coefficients.
4. x, y, and z the space coordinates.
5. The chemical reaction coefficient K. This coefficient is assumed to be linear and constant and characteristic of the pollutant substance and the aquifer material.
6. Other parameters such as the porosity of the aquifer material, temperature, and pressure and the viscosity and density of both the contaminated water and nature groundwater. Entrapped gases also exert an effect.

Every effort must be made to prevent hazardous substances from reaching the aquatic environment. Containing a spill that has entered a waterway is much more difficult than coping with one on land. Most petroleum oils float on water and are highly insoluble and can be retained or directed using floating surface barriers. But many hazardous materials are heavier than water or are soluble, resulting in the contamination of subsurface layers. Insoluble substances that are lighter than water may provide viscosity or corrosive problems that adversely affect floating barriers.

The Ontario Department of the Environment found that it was impossible to effectively stop the spread of oil in the St. Clair river where the current velocity exceeded 0.77 m/sec [2]. Plastic chips, which simulated oil, were found to escape from beneath the retainment booms. In fast-moving currents, booms are effective only in directing floating oil to a collecting device such as a skimmer. At the present time it appears that containment technology is inadequate to cope with oil spills in fast-moving streams. The only course of action is to reduce the chances for spills to occur.

The recovery of insoluble hazardous materials from water may involve a skimming operation or dredging, depending on the specific gravity of the substance. Skimming might involve one of three basic types of operation: vacuum trucks, oleophilic belt skimmers, and inverted belt or impervious skimmers [3]. Vacuum trucks suck up water and oil which are then usually separated by gravity in a tank before the water is returned to the environment. Oleophilic belt skimmers collect oil by allowing water to pass through a polypropylene belt while oil is absorbed on the belt. The belt is squeezed releasing oil into a container. Inverted belt skimmers direct oil into a collection well with an open bottom using an endless conveyor belt. Recovery efficiencies range from 80 to 100%, depending on the density of the pollutant and the specific gravity of the substance.

Skimming operations for nonpetroleum pollutants create other problems. The densities and viscosities of these materials can vary over a wide range making skimming difficult or even impossible. In addition, conventional equipment might be attacked by certain chemicals resulting in the weakening of belts and hoses, or by causing the deterioration of O-rings in pump seals.

Hazardous materials that settle out naturally or through precipitation using an added chemical may be removed from the environment by dredging. Dredging may involve sucking up the pollutant using a hydraulic pipeline ranging in diameter from 20 to 80 cm. A typical system would be a 62-cm diameter suction pipe discharging 1.2 m^3/sec at a discharge velocity of 4.6 m/sec. The effluent is generally pumped into storage bins on the vessel for later transport to land dikes or disposal in open water. Mechanical dippers such as clamshell dredges dump polluted sediments into scows which

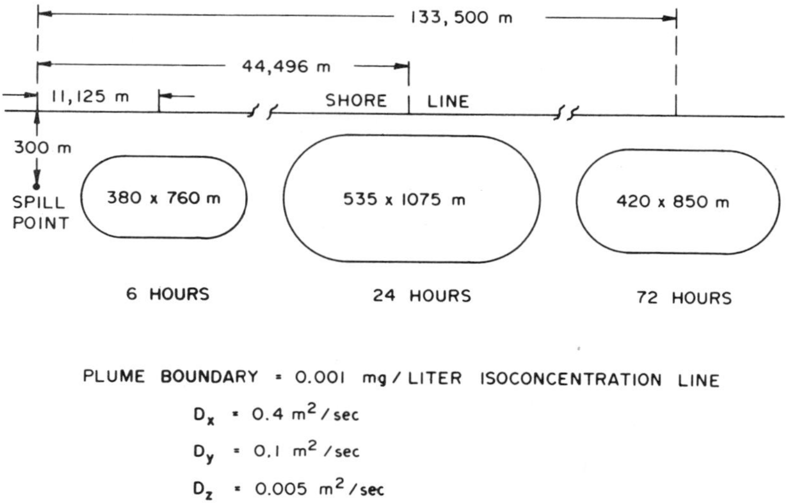

Figure 6-1 Movement of a 3.63-m^3 phenol spill, current moving at 0.51 m/sec. Source: G. W. Dawson, A. J. Shuckrow, and W. H. Swift, Control of Spillage of Hazardous Polluting Substances, U.S. Federal Water Quality Administration, Program No. 1509, Washington, 1970, p. F-4.

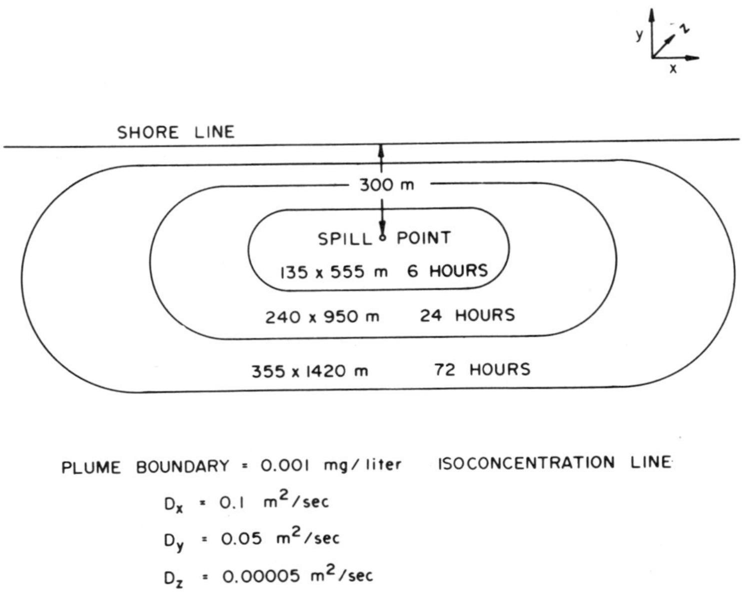

Figure 6-2 Movement of a 3.63-m^3 phenol spill, no current. Source: G. W. Dawson, A. J. Shuckrow, and W. H. Swift, Control of Spillage of Hazardous Polluting Substances, U.S. Federal Water Quality Administration, Program 1509, Washington, 1970, p. F-5.

transport the material to a disposal area. The disposal of contaminated bottom sediments presents a problem. The disposal of contaminated sediments in open water is an undesirable solution.

Dawson [4] illustrated the use of the dispersion model for a spill of 3.63 m^3 of phenol in water. One situation involves a discharge into a body of water where the current velocity is 0.51 m/sec. The plume boundaries of the various isoconcentration lines spread downstream similarly to a smoke plume in air contaminating 133,500 m of shore in 72 hr based on a boundary plume concentration of 0.001 mg/liter. The second situation, involving a spill in static water, contaminates 1420 m of shoreline in 72 hr. The models are summarized in Figures 6-1 and 6-2.

6-1 CONTAINMENT TECHNIQUES

The various methods that have been used to stop leaking containers are summarized below [5]. Normal courses of action involve mechanical repairs to the container, the use of auxiliary containers and retaining devices, and chemical or foam treatments.

1. Rebuilding or reintegrating the container by reassembling, shoring up, reinforcing, pulling separated parts together, and pumping out the damaged container.
2. Building a substitute container by:
    a. Forming dikes
    b. Assembling a knock-down container such as a plastic swimming pool
    c. Using booms, curtains, skimmers, etc.
    d. Digging a pit
    e. Blocking a sewer downstream.
3. Repairing the container by:
    a. Patching with foam or boiler patches
    b. Solidifying or freezing the contents at the leak
    c. Adding something to the substances in the container (internal patch or plug, congealing agent, etc.).

4. Changing the geometric position of the container (up-ending or rotating).
5. Encasing the entire container in a bag.
6. Attaching a collection bag at the place of the leak.
7. Changing the properties of the substance in the container (solidification, gelling, gasification, or burning).
8. Layering or covering over to retard evaporation.
9. Chemical conversion of the substance in the damaged container.

Land spills are more easily dealt with than spills that have reached a waterway. On land, a leaking or ruptured container might be effectively sealed using a quick-setting resinous foam or a patch. Liquid which escapes is directed to a natural or an artificial sump, while the remaining contents of the ruptured tank are pumped into an auxiliary container. Dikes are often constructed to prevent the material from reaching a watercourse. At other times it is possible to seal off a catch-basin and use it as a containment device.

To prevent hazardous substances from permeating into the soil, containment sumps can be lined with plastic sheeting or the walls of the sump can be rendered impermeable by spraying with a quick-setting plastic foam. The construction of an earth dike is time consuming and requires heavy equipment; also the loosened soil tends to absorb the pollutant. Much emphasis is being placed on the development of portable equipment for the construction of foamed plastic dikes [6]. Plastic foams are also being developed that will blanket and deactivate toxic materials.

Oil spill techniques can be used to contain insoluble substances that are lighter than water. This involves the use of skimmers, booms, curtains, slick-lickers, etc. Insoluble substances that are heavier than water may be retained using underwater dikes. Once confined, the substance can be vacuumed or dredged off the bottom.

An interesting method for plugging a ruptured container has been developed by the Rocketdyne Division of Rockwell International Corporation for the U.S. Environmental Protection Agency [7,8]. This method, which is shown in Figure 6-3, consists of a spearhead containing a polyurethane sponge that has been coated with silicone rubber. Urethane foam is introduced into the sponge through openings in a tube forming part of the spearhead. After the injection of the foam base, the sponge swells both inside and out, sealing the ruptured vessel with a hard plug that is held securely in place.

The worst kind of spill involves a toxic, water-soluble cargo that is escaping while the vehicle is submerged in water. Appendix D-1 is a deployment model for an accident of this type. Under ideal conditions, it takes 7 hr for five men to contain any chemical left of the cargo. Under adverse conditions (rough water, bad weather,

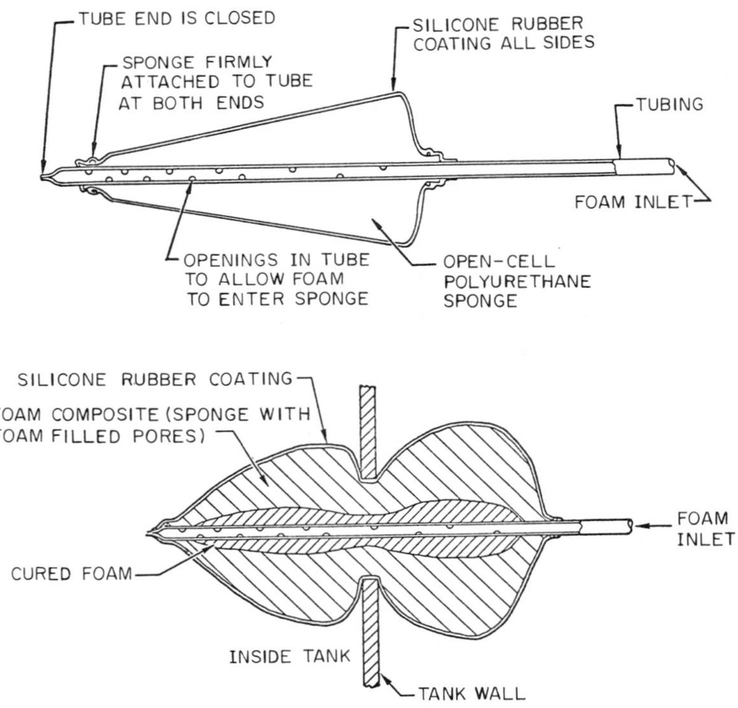

Figure 6-3 Foam composite applicator for plugging leaks. Source: R. C. Mitchell et al., System for Plugging Leaks from Ruptured Containers, Proc. 1974 National Conf. Control of Hazardous Material Spills, San Francisco, August 25-28, 1974, p. 215.

fast current, etc.) a longer time would be required. It is clear from this model and Operation Preparedness [2] that current technology cannot effectively deal with this kind of situation. Large fish kills and extensive environmental damage are inevitable. In fast-moving rivers, the pollutant is carried downstream before the containment systems can be employed. In fact, explosive anchors do not always hold on rocky bottoms of fast-flowing rivers where the force of the current might make it impossible to establish the barrier.

Further assessment of the submerged vehicle situation indicates that stopping the leak is of first priority and independent action must be taken to accomplish this repair. Plastic adhesives exist for this purpose [7]. The plugging operation and the deployment model should be simultaneously coordinated. If the damage is beyond the patching stage, the vehicle might be upended or rolled over, making use of the density difference between the substance and that of water for retainment. Also, pressure from bottom sediments often provides a temporary seal.

Modern technology is rather ineffective in situations of the subject type; thus, prevention of accidents of this kind is a prime factor for adequate control. The following suggestions are offered:

<u>For Highway Transport</u>

1. Highly hazardous cargos, such as those identified herein, should be routed between cities over specifically defined routes.
2. Along defined routes, where the route is adjacent to or crosses a waterway, special arrangements should be made to contain spills in the event of an accident.
3. Special arrangements include:
   a. Shutoff points on ditches adjacent to roadways near waterways,
   b. Lower speed limits in critical areas,
   c. Special truck lanes in critical areas,
   d. Extra-heavy-duty guard rails on bridges and catch basins to direct and retain spills,

e. Mandatory limits on driving hours for drivers hauling dangerous cargos,

f. Mandatory stops under adverse weather or road conditions, and

g. The location of containment hardware depots at strategic points on the route.

<u>For Rail Transport</u>

1. More frequent inspection and repair of tracks which run adjacent to or over waterways,
2. Reduced speed limits in critical areas,
3. More emphasis on the mechanical fitness of rail cars composing a train hauling highly hazardous substances,
4. One-way traffic in critical areas, and
5. The establishment of containment hardware depots at strategic points on the route.

<u>For Marine Transport</u>

1. Highly hazardous cargos should be directed, when in crowded waters, by radio, radar, etc.,
2. Under no conditions should hazardous cargos traverse harbors or inland waterways without the services of a pilot,
3. Further study should be given to identification markings for ships containing highly hazardous cargos,
4. If possible, carriers of highly hazardous materials should tie up under adverse weather conditions and no vessels of this type should be without modern navigational aids, and
5. A review should be made on how clearly bridges, shoals, etc., are marked on crowded waterways.

Environmental damages following a transportation accident can be greatly reduced by accounting for the structural strength of the carrier. Marine carriers are constructed according to the code adopted in 1971 by the U.S. Marine Safety Committee, originally defined by the Intergovernmental Maritime Consultative Organization (IMCO) for ships carrying dangerous cargos [9]. Carriers fall under three classifications:

Type 1. Carriers of highly hazardous cargos requiring maximum construction preparedness:

No portion of the containment system of vessels of this type may be located closer than one-fifth of the vessel's beam from the side of the vessel or a distance equal to one-fifteenth the vessel's beam above the keel. The compartments must withstand a certain prescribed damage rating and the maximum size of an individual compartment is 1250 m^3.

Type 2. Carriers of hazardous cargos requiring significant construction preparedness:

Compartments are required to be a minimum of 76 cm from the sides of the vessel and a distance equal to the vessel's beam divided by fifteen above the keel. Vessels longer than 150 m must have a 2-compartment standard of tank subdivision throughout the length of the ship and individual tank sizes are limited to 3000 m^3. The tanks must pass a certain damage resistance rating.

Type 3. Carriers of cargos requiring moderate construction preparedness:

No separation of cargo containment from the ship's hull is required and a specified level of damage resistance is required.

Special venting and valve facilities are specified for carriers of hazardous materials. Both pressure and vacuum releases are recommended for flammable and toxic cargos having a Reid Vapor Pressure (RVP) less than 172.4 kPa (25 psia). Safety releases are required for nonrefrigerated cargos having an RVP of 172.4 kPa (25 psia) and refrigerated cargos of 276 kPa (40 psia). Manually operated stop valves are considered adequate for cargos not exceeding an RVP of 172.4 kPa at 46° C. Higher pressures require remotely actuated, quick-closing valves.

All carriers of hazardous products must have adequate fire protection and fire-fighting equipment and at least two shutoff locations per compartment. Alarm systems are recommended for highly hazardous materials plus overfill or overflow discharge receivers. Proper color coding and marking systems must be considered.

Barges carrying hazardous and highly hazardous cargos are required to have a void of 7.62 m from the head log to the bow. The spacing between the hull shell and the cargo tanks is 1.22 m for highly hazardous materials and 0.914 m for hazardous materials.

A good deal of environmental damage may be prevented by diking production and storage facilities for hazardous substances. Dikes retain runoffs that might result from spills or contaminated water from a fire. Many new plants and a few modern warehouses have diking facilities. Most facilities have excellent fire-fighting equipment. In the future it is hoped that more and more operations will realize the value of dikes in suppressing environmental damages.

6-2 TREATMENTS FOR SPILLS

Spills exhibit certain characteristics which may include one or more of the following:

1. Gases rising and rapidly dispersing or hovering over the land or water in the spill area
2. Slicks forming on water and harmful substances dissolving into the water from the surface slicks
3. Fine suspensions spreading through the water
4. The rapid dissolving of the substance in water
5. An ooze spreading over benthic sediments

In certain cases, such as the endrin spill at Shawnee Lake, Ohio (Section 3-3), an entire body of water was treated. This is not possible in large bodies of water where excessive dilution has occurred. For effective treatments, it is necessary to contain the pollutant in as highly concentrated a form as possible. A number of water treatment techniques will be discussed in the following section.

6-2.1 Aeration

Many hazardous substances are toxic because they reduce the soluble oxygen content of water. In certain emergencies, it is necessary to maintain adequate dissolved oxygen levels until the harmful substance is removed from the aquatic environment. Equipment is available that can aspirate 4800 vol of air with one vol of water raising the dissolved oxygen content of water from 1 to 10 ppm in a single pass through the mixer [10].

6-2.2 Ozonation and Chlorination

Phenol, cyanides, alcohols, and other substances can be oxidized with strong chemical agents such as ozone, chlorine, and peroxides to levels that are not harmful to the environment. Considerable emphasis is currently being placed on the development of adequate equipment for this purpose. Ozone treatment may be preferable to chlorination since ozone does not leave a harmful residue in the environment. Chlorination is known to produce fish kills.

6-2.3 Neutralization

Sodium bicabonate solutions are commonly used to neutralize spills involving strong acids. Acetic acid is often used for caustic spills.

6-2.4 Precipitation

Precipitating media including sodium carbonate, lime, iron salts, alum, and sodium sulfide solutions are used to precipitate harmful substances such as heavy metal compounds. This is followed by the removal of the precipitated heavy metal compounds which may contaminate the feeding and spawning grounds of fish or cause an upset in the benthic environment. Insoluble residues have been removed by suction techniques.

6-2.5 Carbon Adsorption

Activated carbon is known to effectively adsorb a great variety of organic and inorganic substances. Ziegler and La Fornara [11] have tested the effectiveness of carbon slurries in adsorbing phenol, heavy metal ions, and other compounds (see Appendix D-2 and Table D-2-1). It was found that phenol, soluble mercury, and lead compounds respond favorably to carbon adsorption. Adsorption efficiency relates to the initial concentration of the contaminant; this further emphasizes why containment is so important. For example, in Appendix D-2, Figure D-2-1, about 1814 kg of carbon will remove about 38% of a 454-kg phenol spill when the contaminated water contains 50 ppm phenol. If the treatment were to begin when the concentration of phenol was 1000 ppm, 60% of the original phenol (272 kg) could be removed. The adsorption of phenol on carbon was achieved by agitating contaminated water with a slurry of carbon. Adsorption efficiency is increased by increasing the ratio of carbon to pollutant concentration. This is shown in Table D-2-1 of Appendix D-2. For example, with a ratio of 5:1 of carbon to lead, about 13% of the lead is removed in a 107-ppm solution. At a ratio of 50:1, the treatment removes 84% of the lead from the same solution.

6-2.6 Ion Exchange Resins

Ion exchange resins are effective for removing a large number of toxic ionic substances from water. Included in this group are acids, bases, salts of heavy metals, and other ionic substances. Acids and bases can be effectively neutralized and ions such as Cd^{++}, Pb^{++}, Hg^{++}, etc., are removed from the ambient water and replaced by Na^{+} ions.

The degree of sorption for a particular ion depends on the concentration and type of competing ions and the temperature of the system. Sorption may be accomplished by passing contaminated water through a column or by the agitation of a slurry of contaminated water with an ion exchange resin. Research projects are in progress to test the effectiveness of adding ion exchange resins directly into contaminated water [12].

Table 6-1 summarizes treatments that may be attempted for the highly hazardous substances defined in this study.

Table 6-1

Deactivation Techniques for Hazardous Substances

Highly Toxic Substances	Treatment
Mercury fungicides	Carbon adsorption
Pesticides, polychlorinated biphenyls	Carbon adsorption and the removal of contaminated sediments to a disposal area
Mercuric salts	Dimercaprol, a water-soluble mercaptan, is recommended as a chelating agent for mercury; sodium carbonate will precipitate soluble mercury as mercuric carbonate, which can then be removed from bottom sediments
Cadmium salts	Precipitation with sodium carbonate and/or the use of chelating agents
Potassium cyanide, hydrogen chloride	Precipitation of CN^- with ferric salts or chlorination; acidic pH values must be avoided to prevent the release of HCN gas; slaked lime will raise the pH
Acrylonitrile	Decomposes in water with the release of HCN gas; slaked lime, ferric salts, and/or chlorination treatments may be tried
Tetraethyl lead solutions	Skimming followed by carbon adsorption and possibly chelating agents such as calcium ethylenediamine tetraacetate

Table 6-1 (continued)

Highly Toxic Substances	Treatment
Toxic Gases	
Phosgene	Prompt evacuation appears to be the only solution
Hydrogen sulfide	Ferric salts are generally tried when the gas is being released from an aquatic environment
Chlorine	Caustic solutions, powdered carbon, and foam and water curtains have been tried; regardless of what is done, if the runoff reaches the aquatic environment, large fish kills are inevitable
Corrosive Liquids	
Acids in general	Neutralization with slaked lime or sodium bicarbonate followed by dilution of the corresponding salt in water
Others	
Phenol	When spills are relatively small and confined, carbon adsorption is effective; for large spills, dilution may be the only answer (at a pH of 6, phenol forms water-soluble salts)
Ethylenimine	Neutralization with acetic acid and adsorption of the neutral product with carbon

Source: G. W. Dawson, A. J. Shuckrow, and W. H. Swift, Control of Spillage of Hazardous Polluting Substances, U.S. Federal Water Quality Administration, Program No. 1509, Washington, 1970.

6-3 MOBILE TREATMENT SYSTEMS

A system of preparedness requires that facilities for the containment and treatment of spills shall have a high degree of mobility. The components in a mobile treatment system should contain reaction, flocculation, and sediment tanks, filters, and adsorption tanks.

The entire system might be contained on a 13.7-m trailer. Treatment and waste tanks may be of a rigid or collapsible nature with easily assembled plastic and stainless steel connecting lines. The schematic of such a system is shown in Figure 6-4.

Contaminated water is pumped to the initial treatment tanks where deactivating reactions and sedimentation can occur. The treated water is then filtered and passed through carbon beds to remove undesirable trace materials such as phenol, organometallic substances, pesticides, etc. The effluent is aerated before returning to the environment. Contaminants retained in the sediment tank and filters and material washed from the carbon beds are stored in separate tanks for later disposal.

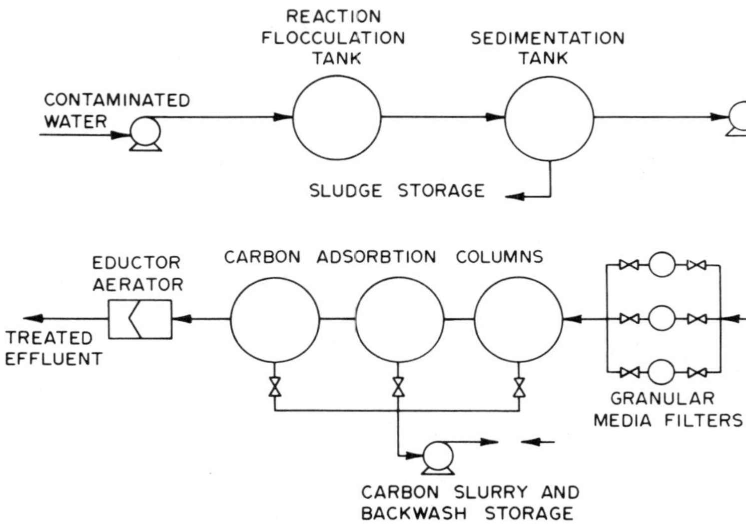

HYDRAULIC CAPACITY	759 liter/min (200 U.S. gal/min)
CARBON CAPACITY	8172 kg (18,000 lb)
GRANULAR FILTER AREA	2.7 m^2 (30 ft^2)

Figure 6-4 A mobile treatment system. Source: D. G. Mason, M. K. Gaptu, and R. C. Scholz, A Mobile Multipurpose Treatment System for Processing Hazardous Material Contaminated Waters, Rex Chainbelt Inc., Proc. 1972 National Conf. Control of Hazardous Material Spills, Houston, Texas, March 21-23, 1972, Lib. Cong. Cat. No. 72-77216, p. 154.

It might be pointed out that a mobile system of this type is very heavy, especially if the carbon is moistened to increase initial adsorption efficiency. A weight of 45,400 kg may be expected. Since not all overpasses and bridges are designed for such a weight, some route planning is necessary in utilizing the mobile concept.

The Rexnord Corporation [13] designed a system for the removal of creosote from the bottom sediments of the Menomonee river in Milwaukee. This system is outlined in Figure 6-5. A "river sweeper" was used to vacuum contaminated mud from the bottom of the river. The vacuumed mud was pumped to a beach cleaner previously developed by the Hazardous Spill Technology Branch of the Industrial Waste Treatment Research Laboratory of the U.S. Environmental Protection Agency, to clean oil from beach sand. The creosote-mud slurry was air blown in the beach cleaner causing creosote to separate from the mud and rise to the surface with the air bubbles. After separating the froth containing the creosote, the mud slurry was pumped to a flocculation tank where ferric chloride was added to hasten

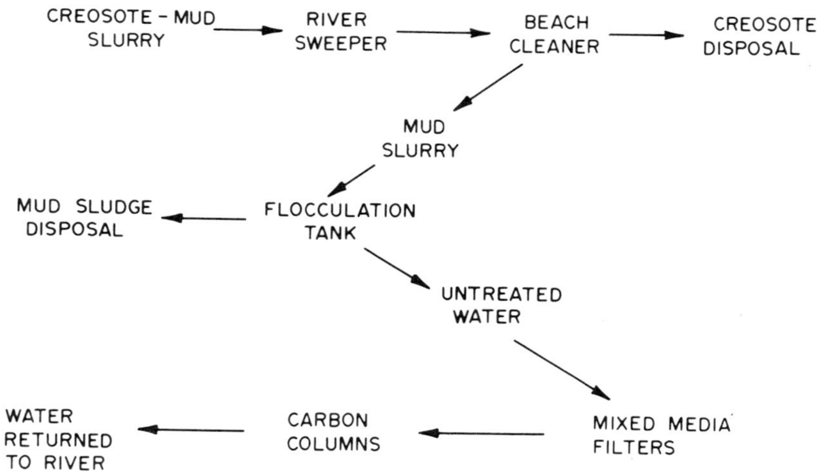

Figure 6-5 Creosote removal system (Rexnord Corporation). Source: J. P. La Fornara and I. Wilder, Solution of the Hazardous Material Spill Problem in the Little Menomonee River, Proc. 1974 National Conf. Control of Hazardous Material Spills, San Francisco, August 25-28, 1974, p. 203.

the settling of the mud. The settled mud was forwarded to a disposal area, while the supernatant water was passed through mixed-media filters and carbon columns to remove the last traces of creosote. For a total of 435,600 liters of slurry processed, 64,300 liters of mud was trucked to a landfill and 408,800 liters of treated water was returned to the river. The average creosote concentration of the treated effluent water was 2.3 mg/liter. In this process, 98% of the creosote and 99% of the suspended solids were removed from the water. The removal effectiveness of the Rexnord system is outlined in Table 6-2.

Table 6-2

Creosote Removal Effectiveness — Rexnord Treatment System

	Beach Cleaner	Flocculation Tank	Mixed-Media Filters	Carbon Columns	Total System
Influent Creosote content, mg/liter	1800	370	21	8.1	1800
Effluent Creosote content, mg/liter	370	21	8.1	2.3	2.3
Percentage of remaining creosote removed	79	94	61	72	--
Percentage of total creosote removed	79	19	0.7	0.3	99+
Cumulative percentage of creosote removed	79	98	99	99+	99+

Source: J. P. Fornara and I. Wilder, Solution of the Hazardous Material Spill Problem in the Little Menomonee River, Proc. 1974 National Conf. Control of Hazardous Material Spills, San Francisco, August 25-28, 1974, p. 204.

REFERENCES

1. A. A. Metry, Prediction, Control and Recovery of Hazardous Substances Migrating in Subsurface Waters, Proc. 1974 National Conf. Control of Hazardous Material Spills, San Francisco, August 25-28, 1974, pp. 262-272.
2. Presentation by the Water Management Branch of the Ontario Dept. of the Environment to the Chemical Institute of Canada, Sarnia, Ont., Oct. 1, 1974.
3. E. E. Johanson and S. P. Bowen, The Recovery and Processing of Hazardous Spills in Water, Proc. 1974 National Conf. Control of Hazardous Material Spills, San Francisco, August 25-28, 1974, pp. 188-193.
4. G. W. Dawson, A. J. Shuckrow, and W. H. Swift, Control of Spillage of Hazardous Polluting Substances, U.S. Federal Water Quality Administration, Program No. 1509, Washington, 1970, pp. F-4 and F-5.
5. I. Wilder and J. E. Brugger, Present and Future Technology Requirements for the Containment of Hazardous Material Spills, Proc. 1972 National Conf. Control of Hazardous Material Spills, Houston, Texas, March 21-23, 1972.
6. R. H. Hiltz, M. D. Marshall, and J. V. Friel, The Physical Containment of Land Spills by a Foam Diking System, MSA Research Corp., EPA Contract No. 68-01-0100, Proc. 1972 National Conf. Control of Hazardous Material Spills, Houston, Texas, March 21-23, 1972, Library of Congress Cat. No. 72-77216.
7. R. C. Mitchell, M. Kirch, E. L. Hamermesh, and J. E. Sinor, Methods for Plugging Leaking Chemical Containers, Rocketdyne-North American Rockwell Corp., Proc. 1972 National Conf. Control of Hazardous Material Spills, Houston, Texas, March 21-23, 1972, Library of Congress Cat. No. 72-77216.
8. R. C. Mitchell et al., System for Plugging Leaks from Ruptured Containers, Proc. 1974 National Conf. Control of Hazardous Material Spills, San Francisco, August 25-28, 1974, pp. 212-216.
9. R. T. Luckritz, Hazardous Materials Spills Prevention in the Bulk Marine Carriage of Dangerous Cargoes, Proc. 1974 National Conf. Control of Hazardous Material Spills, San Francisco, August 25-28, 1974, pp. 18-24.
10. R. G. Sanders, S. R. Rich, and T. G. Pantazelos, A Short Contact Time Physical Chemical Treatment System for Hazardous Material Contaminated Waters, Proc. 1972 National Conf. Control of Hazardous Material Spills, Houston, Texas, March 21-23, 1972.

11. R. C. Ziegler and J. P. La Fornara, In Situ Treatment Methods for Hazardous Material Spills, Proc. 1972 National Conf. Control of Hazardous Material Spills, Houston, Texas, March 21-23, 1972.

12. A. J. Shuckrow, B. W. Mercer, and G. W. Dawson, The Application of Sorption Process for In Situ Treatment of Hazardous Material Spills, Proc. 1972 National Conf. Control of Hazardous Material Spills, Houston, Texas, March 21-23, 1972, Library of Congress Cat. No. 72-77216.

13. J. P. La Fornara and I. Wilder, Solution of the Hazardous Material Spill Problem in the Little Menomonee River, Proc. 1974 National Conf. Control of Hazardous Material Spills, San Francisco, August 25-28, 1974, pp. 202-207.

Chapter 7

RADIOISOTOPES AND THE ENVIRONMENT

There is increasing concern over the pollution of the biosphere by radioisotopes. Power demands are resulting in the construction of an increasing number of nuclear power plants throughout the world. Merun [1] estimates that there will be a growth in world nuclear capacity from 2.02×10^4 MW(e) in 1970 to 1.33×10^6 MW(e) in 1990. At the present time emissions of radioisotopes to the environment from all sources are well below natural levels but the expanding use of nuclear energy will undoubtedly elevate these levels, and there is increasing awareness that power plant accidents in the future may release excessive radioisotopes into populated areas.

Surprisingly, more radioisotopes are released to the environment by the combustion of fossil fuels than by the nuclear industry. In 1972, the National Academy of Engineering [2] reported that emissions from the combustion of fossil fuels are equal to or greater than the emissions from currently operating nuclear power plants. A typical coal contains 2.0 ppm thorium-232 and 1.1 ppm uranium isotopes. If a thermal electric plant burns 2.1×10^6 metric tons per year of coal with a collection efficiency of 97.5% fly-ash removal, the release of radioisotopes is equivalent to 17.2 mCi* of radium-226 and 10.8 mCi of radon-228. Radium-226 and 228 are daughter products of uranium-235 and thorium-232. This is equivalent to

*See Appendix E for definitions of atomic terms.

10^4 Ci of krypton-85 or 10 Ci of iodine-135, which are two gaseous radioisotopes of concern in nuclear power plant stack emissions. Equivalent oil emissions for the same amount of thermal energy is much lower, i.e., 0.5 mCi equivalency based on radium-226 and -228.

The annual exposure of people in the United States to natural radiation ranges between 100 mrem at sea level to 170 mrem at Denver, Colorado [3]. In some areas in Brazil and India natural background levels are 10 times higher. On certain occasions the environment is exposed to fairly intensive surges of solar radiation. For example, on February 23, 1956, it was reported that over 100 mrem/hr of radiation was detected at 10,670 m altitude from solar flare activity [4].

Man-made sources of radiation include emissions from nuclear power plants, releases during the processing of spent reactor fuels, weapon testing, accidental discharges, releases from propulsion devices, and miscellaneous sources such as the discharge of research isotopes into sewers. Tables 7-1 and 7-2 list radioisotopes with their sources and body effects.

Table 7-1

Sources of Radioisotopes

	Nuclear Reactor	Fuel Reprocessing	Fabrication Facilities for Fuels[a]	Aerospace
H-3	x	x		
C-14	x			
Noble gases				
Xe-133, 135, 138	x	x		
Kr-85, 87, 88	x	x		
Halogens				
I-129		x		
I-131	x	x		
Bone seekers				
Pu-238, 239	x	x		x
U and Th		x	x	x
Sr-89, 90	x	x		
Ra-226			x	
Po-210			x	

Table 7-1 (continued)

	Nuclear Reactor	Fuel Reprocessing	Fabrication Facilities for Fuels[a]	Aerospace
Others				
Zr-95	x	x		
Co-58, 60	x	x		
Cs-134, 137	x	x		
Ce-144	x	x		
Ru-106	x	x		
Zn-65	x			

Source: National Academy of Engineering, Engineering for Resolution of the Energy-Environment Dilemma, Printing and Publishing Office, National Academy of Sciences, Washington, 1972, Library of Congress Cat. No. 79-186370.
[a]Mines, mills, etc.

The Atomic Energy Commission has placed strict limits on air and water discharges from nuclear power plants. Air emissions under normal operating conditions are such that an exposure of 0.5 rem will not be exceeded even if one were exposed continuously for one year at the point of highest concentration at ground level [5]. Water effluents are maintained at a level suitable for human consumption without ill effects at the point of discharge.

On several occasions accidents have occurred with atomic reactors releasing serious amounts of radioisotopes to the environment. An accident occurred at Chalk River, Ont., on December 12, 1952 [6] resulting in 100 people receiving excessive dosages of radiation and the emission of radionuclides over an unpopulated area. This was an unusual accident involving a plexiglas sight glass that was left incidentally in a research reactor. The plexiglas melted, plugging circulation lines resulting in the overheating and the failure of the reactor. Sight glasses are now marked with stripes to avoid future similar mishaps.

Another accident occurred at Windscale, England, spreading radioactivity over an area of 518 km^2. Sufficient quantities of radioactivity were found on vegetation to terminate milk production

Table 7-2

Radioisotopes, Body Absorption Characteristics, and Limits

Isotope	Half-Life	Radiation Emitted	Absorption via Lungs	Elimination from Body	Localization in Body	Maximum Amount Permitted in Body[a]
P-32	143 days	β	Excellent	Few weeks	Bone	10 μCi
Sr-89	55 days	β	Excellent	Slow	Bone	1.0 μCi
Sr-90	25 years	β	Excellent	Very slow	Bone	0.5 μCi
I-131	8 days	β,γ	Complete	1 Month	Thyroid	0.1 μCi
Po-210	140 days	α,γ	Good	Slow	Kidneys	0.005 μCi
Ra-226	1590 years	α,γ	Good	Very slow	Bone	0.1 μg
U-238	4.6×10^9 years	α	Poor	Slow	Lungs, kidneys	0.6 μg
Pu-239	2.4×10^4 years	α	Very poor	Very slow	Lungs	0.1 μg
Tritium	12.26 years	β	Good	Slow	Total body	2000 μC

Source: RCA Service Company, Atomic Radiation, Camden, New Jersey, 1967.

[a]Personnel exposure, multiply by 10 for point of minimal damage. For large centers of population, divide by 100.

for from three to six weeks following the accident. This accident was caused by the failure of a graphite control rod in an air-cooled reactor.

In yet another case in Idaho on January 3, 1961, Iodine-131 was deposited downwind for 161 km. This particular reactor contained enriched fuel. Cases such as those mentioned above clearly indicate that serious problems can occur with atomic reactors and the environment is vulnerable.

As mentioned previously, sources other than nuclear reactors release radiation to the environment. During the testing of weapons radioactive dust is suspended in the upper troposphere. This dust is transported over a wide area releasing radioisotopes such as plutonium-239, strontium-90, cesium-137, and iodine-131, which are reported to have added from 10 to 15% to the natural background radiation throughout the world [7]. In June 1962, a device containing plutonium-238 and curium-244 was put in orbit. This satellite burned up on reentry over the Indian Ocean in 1964 leaving detectable traces of plutonium in the atmosphere. The processing of spent nuclear fuel rods for uranium-238 and plutonium-239 releases fissionable products such as krypton-85, xenon-133, and iodine-131, and also tritium to the atmosphere.

On June 4, 1971, the Atomic Energy Commission proposed the following guidelines for the design of nuclear power plants [2].

1. Reasonable assurance must be provided that annual exposures to individuals near the boundary of nuclear power plant sites will be less than 5% of average exposures from natural background levels (about 100 mrem).
2. Reasonable assurance must be provided that the annual exposure to sizable population groups from releases of radioisotopes from nuclear power plants in the foreseeable future will be less than 1% of exposures from natural backgrounds.

It is generally accepted by the AEC and other agencies that no single dose per year for any individual shall exceed 0.5 rem, and population groups shall not be subjected to levels greater than

Table 7-3

Permissible Limits of Radioisotopes

Isotope	Air	Water
P-32	2×10^{-8} μC_i/ml	2×10^{-4} μCi/ml
Sr-90	1×10^{-10} μC_i/ml	2×10^{-6} μCi/ml
I-131	1×10^{-9} μC_i/ml	1×10^{-5} μCi/ml
Ra-226	4×10^{-12} μC_i/ml	4×10^{-8} μCi/ml
U-238		
Soluble	6×10^{-9} μg/ml	2×10^{-3} μg/ml
Insoluble	2.5×10^{-11} μg/ml	
Pu-239		
Soluble	5×10^{-12} μg/ml	4×10^{-6} μg/ml
Insoluble	2.5×10^{-11} μg/ml	

Source: RCA Service Company, Atomic Radiation, Camden, New Jersey, 1967.

0.170 rem per year [2]. Permissible amounts for various radioisotopes in air and water are listed in Table 7-3.

Two nuclear fuels occur in nature; these are ores of uranium and thorium. Most uranium-bearing minerals contain only a small concentration of uranium. The main uranium oxide ores are pitchblende and carnotite which are found in the Great Bear Lake region in Canada, the Joachimsthal district in Czechoslovakia, the Colorado plateau in the United States, and Zaire. Uranium ores contain three uranium isotopes: U-238 (99.28%), U-235 (0.71%), and U-234 (0.00918%). Thorium deposits are found in India, Brazil, Ceylon, Tasmania, Nigeria, the Ural mountains, and Scandinavia. Of the three uranium isotopes, only U-235 can sustain a chain reaction.

The fission of U-235 produces a number of radioisotopes including noble gases, halogens, and other elements. One possible fission process is as follows:

$$\xrightarrow{\text{slow neutron impact}} \quad {}_{92}U^{235} \rightarrow {}_{56}Ba^{144} + {}_{36}Kr^{90} + 3\,{}_{0}n^{1} + 200 \text{ MeV}$$

The greatest portion of the released energy is transported in the fission products (barium and krypton) and accounts for 167 MeV. Neutron energies account for about 6 MeV and gamma radiation for decay processes. During the fission of 1 kg of U-235, 0.9990 kg of fission products are produced and 0.0010 kg is converted to energy. The energy release from 1 kg of U-235 is equivalent to 2.51×10^7 kW-hr of power. Various side reactions that occur in reactors are shown in Appendix E. Breeder reactions are important for the production of raw materials for weapons.

The radiation emitted by radioisotopes can destroy cell tissue by producing ions when the radiation passes through an organism. This process is shown in Figure 7-1. Ion pairs are produced during impacts with cell tissue which leads to a cascading effect involving secondary ionizations. The total ionization produced depends on the nature and energy of the radiation. Alpha radiation produces any-

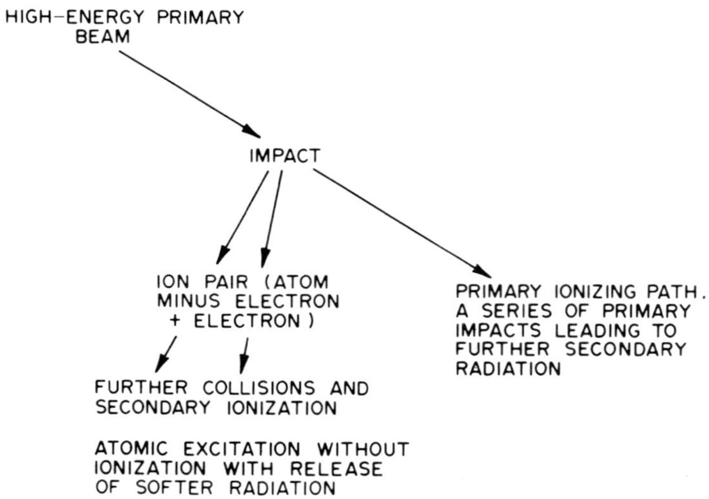

Figure 7-1 The process of ionization by radiation. Degree of ionization depends on mass, charge, and velocity of primary beam, and collision angles. On the average, the formation of a single ion pair in air at STP requires the expenditure of about 32.5 eV of energy.

where from 30 to 100,000 ion pairs per cm of dry air as compared to about 100 pairs for beta radiation. Gamma radiation is much more penetrating than either alpha or beta radiation but produces fewer ion pairs.

Certain cells are more susceptible to radiation [7,8] than others, as shown in the following list [7,9].

Increasing sensitivity

1. White blood cells (lymphocytes) formed in the spleen and lymph nodes
2. White blood cells (granulocytes) formed in bone marrow
3. Basal cells, specialized cells of the gonads, bone marrow; skin and alimentary canal
4. Alveolar cells which absorb oxygen and release carbon dioxide from the blood
5. Bile duct cells
6. Cells of the tubules of the kidneys
7. Endothelial cells that line the closed cavities of the body such as the heart and blood vessels
8. Connective tissue cells
9. Muscle cells
10. Bone cells
11. Nerve cells

In general, cells are more susceptible at certain stages of division and the higher the metabolic rate in a cell, the lower its resistance to radiation. Cells involved in the reproductive processes are especially susceptible to damage from radiation which promotes increased mutations. Permanent damage to cells is caused by the formation of free radicals during ionization, as shown in Figure 7-2. The free radicals induce the formation of abnormal linkages that result in mutations and usually cell death. The following list of radioisotopes shows their preferences for human tissue (see also Table 7-1).

CHEMICAL CHANGES IN WATER

$$H_2O \xrightarrow[\infty,\ \beta \text{ OR } \gamma]{\text{IONIZING RADIATION}} H_2O^+ + e^-$$

$$H_2O + e^- \longrightarrow H_2O^-$$

$$H_2O^+ \longrightarrow H^+ + OH\cdot \qquad \text{NET FORMATION,}$$

$$H_2O^- \longrightarrow OH^- + H\cdot \qquad \text{TWO FREE RADICALS}$$

$$H^+ + \dot{O}H^- \longrightarrow H_2O$$

FREE RADICAL REACTIONS

$$H\cdot + OH\cdot \longrightarrow H_2O$$

$$H\cdot + H\cdot \longrightarrow H_2$$

$$OH\cdot + OH\cdot \longrightarrow H_2O_2$$

PROTEIN $\quad -SH + -SH \xrightarrow{2OH\cdot} -S-S- + 2H_2O$

CELL HYDROCARBONS

$$RH + OH\cdot \longrightarrow R\cdot + H_2O$$

$$R\cdot + O_2 \longrightarrow RO_2\cdot$$

CELL DAMAGE

$$RH \xrightarrow{\text{IRRADIATION}} RH^+ + e^-$$

$$RH^+ \longrightarrow R\cdot + H^+$$

$$R\cdot + R\cdot \longrightarrow R-R$$

PROTEIN (ABOVE) WITH SULFHYDRYL GROUPS
CELL HYDROCARBONS (ABOVE)

Figure 7-2 The effects of radiation on matter.

Isotope preferences for human tissues:

Lungs

Radon-222	Polonium-239
Uranium-238	Nickel-63
Plutonium-239	

Kidneys

Uranium-238	Arsenic-76
Chromium-51	Rhodium-105, 106
Manganese-56	Gold-198
Germanium-71	

Liver

Manganese-56	Copper-64
Nickel-59	Silver-105, 109, 111
Cobalt-60	Cadmium-109

Bone

Strontium-89, 90	Yttrium-90, 91
Carbon-14	Tin-113
Beryllium-7	Barium-140
Fluorine-18	Tungsten-185
Phosphorus-32	Lead-203
Calcium-45	Radium-226
Vanadium-48	Uranium-233
Zinc-65	Plutonium-239

Tritium, with a half-life of 12.3 years, may be released to the environment during nuclear reactions. The major part of the tritium produced in nuclear-electric plants is released in the liquid waste effluents [3]. It is estimated that about 6% of the total tritium released from a pressurized water reactor is in the form of atmospheric emissions. Heavy water losses from reactors approach several kilograms per day, and from 1000 to 10,000 Ci of tritium per month are released to the environment [3]. Based on past experience, the Douglas Point Nuclear Station in Ontario releases 4 kg per day of heavy water, which could indicate a release of 15,000 Ci of tritium per year to the environment [9]. The largest localized concentrations of tritium to the environment are suspected to be in the vicinity of reprocessing plants for spent fuel elements.

Canadian nuclear power plants are fueled with low-cost, natural uranium bundles. It may be argued that natural uranium reactors of the Candu type are safer than fast neutron reactors, many of which are used in the United States. Fast reactors with enriched fuels are somewhat more delicate to control and are prone to overheat. Design features of the Candu heavy water reactors with their cheap natural fuel preparations are gaining world renown even though initial investments are high.

The Candu Station of the Ontario Hydro Electric Commission at Pickering, Ontario [10] is shown in Figure 7-3. This plant has four pressure-tube reactor units that are moderated and cooled by heavy water. The heavy water cooling cycle exchanges heat with light water

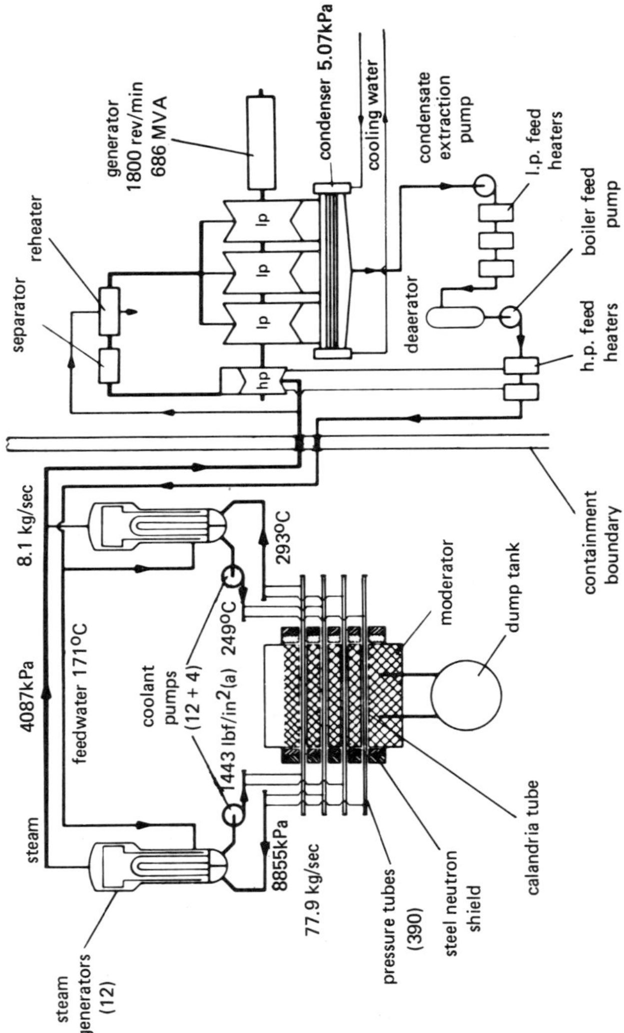
steam
4087kPa
8.1 kg/sec
feedwater 171°C
steam generators (12)
coolant pumps (12 + 4)
1443 lbf/in²(a)
249°C
293°C
8855kPa
77.9 kg/sec
pressure tubes (390)
steel neutron shield
calandria tube
moderator
dump tank
containment boundary
separator
reheater
hp
lp
lp
lp
generator 1800 rev/min 686 MVA
condenser 5.07kPa
cooling water
condensate extraction pump
l.p. feed heaters
deaerator
boiler feed pump
h.p. feed heaters

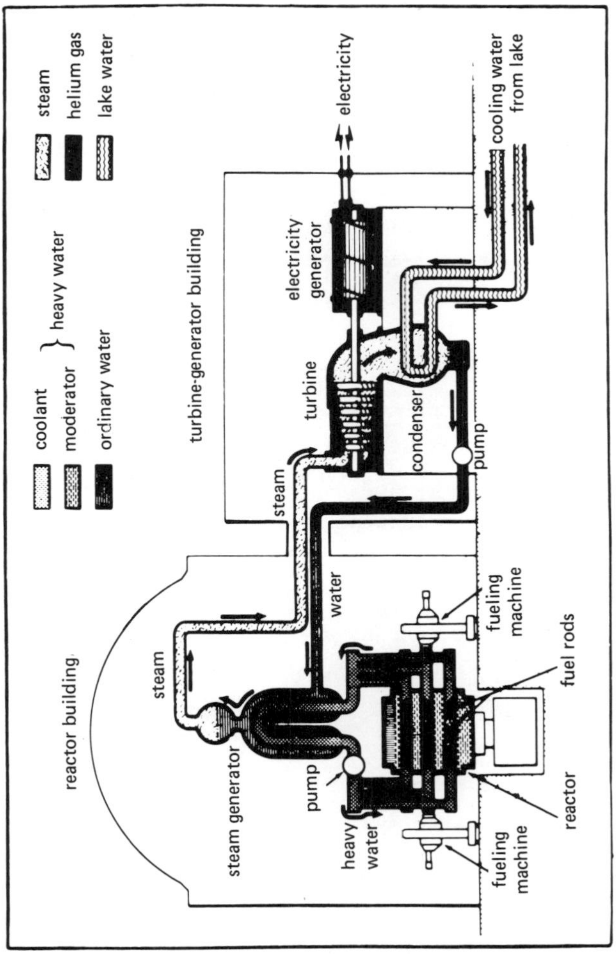

Figure 7-3 The Candu plant at Pickering, Ontario. Source: R. N. Moore, Nuclear Power, Cambridge University Press, London, 1971.

which generates steam to run the turbines. Each reactor unit produces approximately 500 MW(e) and is designed to be refueled while on load at a rate of up to seven bundles per day per reactor. Refueling is a complex operation which is totally automated and digitalized with alternate programming and manual facilities in event of emergencies. Should an accident occur in the reactor section of the plant, a negative pressure system has been designed for each reactor section to retain all radioactive materials within the reactor building. The reinforced concrete walls of the reactor building are designed to stand 41.4 kPa (6 psig) internal pressure, and a large water tank of 991 m^3 capacity is available to discharge water into the reactor area to reduce heat-induced pressure. Air in the reactor room is recirculated to remove traces of heavy water and radioactive gases. During start-up, light water condensate is continuously monitored for heavy water leaks by infrared absorption analysis, and when the reactor is on stream, monitoring is continued for tritium oxide rather than heavy water.

The AEC prescribes a stack discharge limit of 7×10^5 μCi/sec from a nuclear power plant. The principal radioactive gases in stack effluents include isotopes of krypton-85, 87, and 88, tritium, and xenon-133, 135, and 138. Radioactive constituents of liquid wastes include tritium, cobalt-58, strontium-89 and 90, iodine-131, cesium-134 and 137, barium-140, and cerium-144. The permissible limit set by the AEC is 1×10^{-7} μCi/ml for an unidentified mixture containing no iodine-129, radium-226, or radon-228, and 3×10^{-7} μCi per ml for effluents containing strontium-90 and iodine-131.

Typical off-gas and liquid waste systems for nuclear power plants are shown in Figures 7-4 and 7-5. In off-gas systems, noncondensable gases are drawn from the main condenser through steam jet air ejectors to delay and air filter systems, and are then exhausted up the stack. Wastes are segregated according to their chemical and physical properties in liquid waste systems. The liquid wastes are filtered and directed to a discharge canal, while solids are centrifuged and packaged for permanent storage. The Candu-Pickering plant

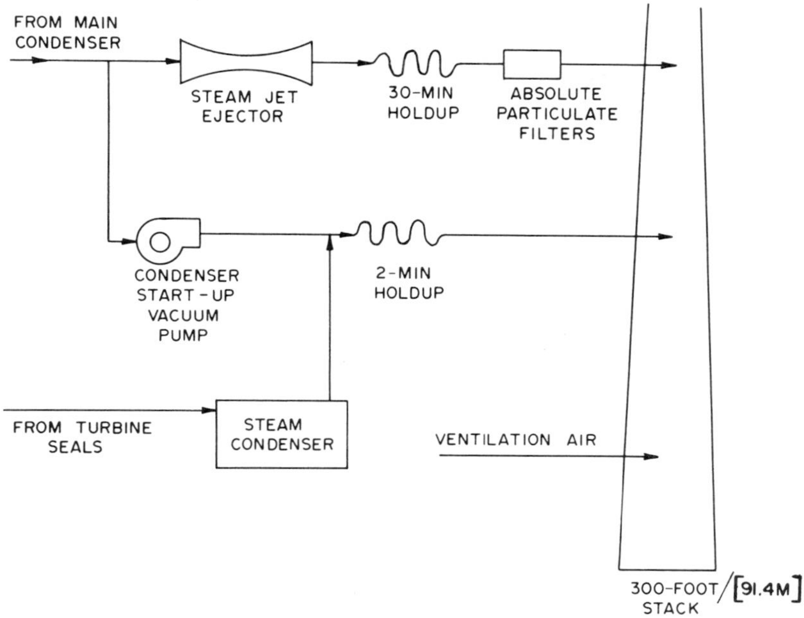

Figure 7-4 Off-gas system for a nuclear power plant. Source: National Academy of Engineering, Engineering for Resolution of the Energy-Environment Dilemma, Printing and Publishing Office, National Academy of Sciences, Washington, 1972, p. 173.

has underwater storage capacity for 50 years of normal operations; all spent fuel bundles are sent to storage.

The National Academy of Sciences [2] estimates that in the near future cryogenic enrichment techniques will be used for the removal of gaseous radioisotopes from power plant effluents. Off-gases might be compressed to about 304 kPa, then cooled to 4.4° C to remove water vapor. After further cooling to -179° C (94 K) with liquid nitrogen, isotopes of krypton and xenon are removed. The processes involved include absorption using a solvent such as a fluorocarbon followed by stripping and fractionation. A technique for the removal of gaseous tritium from effluent gases by the catalytic conversion of tritium to tritium oxide is being investigated. Chemical separation techniques are being tried for the removal of radioisotopes from liquid wastes followed by physical removal techniques.

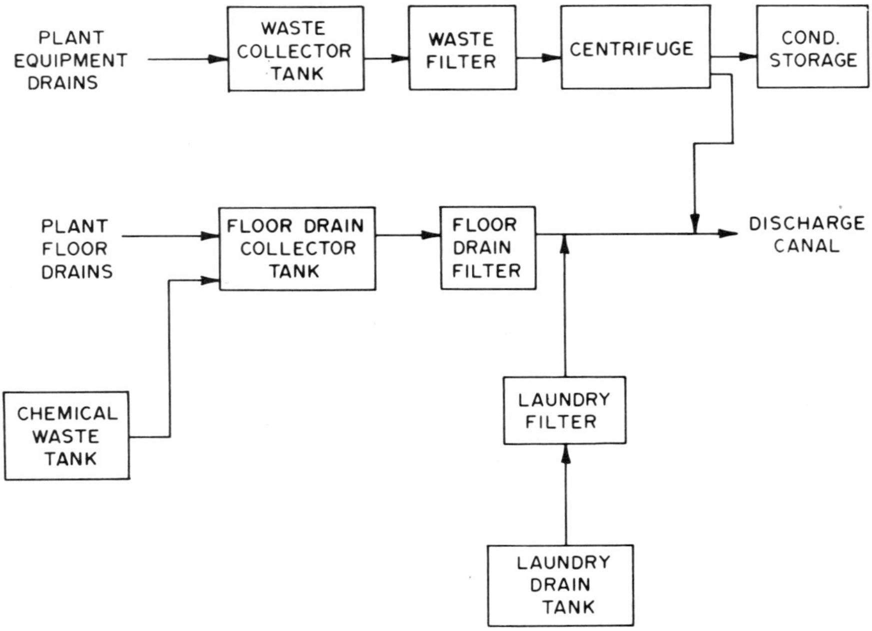

Figure 7-5 Liquid effluent system, nuclear plant. Source: National Academy of Engineering, Engineering for Resolution of the Energy-Environment Dilemma, Printing and Publishing Office, National Academy of Sciences, Washington, 1972, p. 174.

One can only speculate on the role of nuclear power in the generation of electric power over the next 50 years. Coal and hydro plants will probably continue to be prime power sources for years to come. Nuclear plants of the Candu type appear to have more appeal on a world basis than fast neutron reactors but represent higher initial investments. Heavy water reactors operate on natural fuel which is much less costly to prepare. Although thermal and hydro plants may be prime power sources, it is expected that nuclear plants will handle surge requirements. Local peak requirements will be further supplemented by gas turbine power using gasified fuels or sewer gases as fuel. Nuclear fusion is being developed; however, it is questionable when this process will reach commercial importance. When perfected, fusion reactors will probably replace fission and thermal reactors since the fuel supply for fusion reactors is essentially inexhaustible. There is already concern that high-assay uranium ores may be exhausted in the next century; it will then be necessary to mine poorer grade ores which will raise the cost of fuel. Very likely, spent fuels will be reprocessed which will produce increased radioemissions. Also, it is speculated that enriched fuels and hotter reactors will be necessary in the next century. Whatever direction is taken, an improved environment is a requirement.

REFERENCES

1. H. B. Merun, Nuclear Energy, Its Growth and Impact, Canadian Nuclear Association Economic Development Committee Report 71-CNA-301, June 1971.

2. National Academy of Engineering, Engineering for Resolution of the Energy-Environment Dilemma, Printing and Publishing Office, Natl. Acad. of Sciences, Washington, 1972. Lib. of Congress Cat. No. 79-186370.

3. U.S. Congress, Hearings Before the Joint Committee on Atomic Energy, 1969, Environmental Effects of Producing Electric Power, U.S. Govt. Print. Off., Washington, 1969.

4. H. J. Schaefer, Radiation Protection: Recommendations of the International Commission on Radiological Protection, Publication No. 9, Pergamon, New York, 1966.

5. S. Miner, Air Pollution Aspects of Radioactive Substances, U.S. Dept. Commerce, National Bureau of Standards, PB 188092, Litton Industries, Inc., Bethesda, Maryland, 1969.

6. W. B. Lewis, The Accident to the NRX Reactor on December 12, 1952, AECL 232, Atomic Energy of Canada, Ltd., Chalk River, Ont., 1953.

7. RCA Service Company, Atomic Radiation, Camden, N.J., 1967.

8. P. Alexander, Atomic Radiation and Life, Penguin Books, New York, 1965.

9. G. H. Williams, Douglas Point Generating Station Commissioning Heavy Water Power Reactors, Proc. 1967 Intern. Atomic Energy, Vienna, 1967.

10. R. V. Moore, Nuclear Power, Cambridge University Press, London, 1971.

PART 2

A RATING SYSTEM FOR COMMUNITIES EXPOSED TO HAZARDOUS MATERIALS

INTRODUCTION

The production facilities for hazardous materials are often far removed from market locations. In Canada, chemicals manufactured in the Toronto-Hamilton area may be transported to the east or west coast of the Dominion. Many United States production sites for toxic substances such as pesticides are located in the eastern half of the country. High-production areas include New York-New Jersey, Baltimore-Wilmington, West Virginia, Chicago, Cleveland-Detroit, St. Louis-Kansas City, Houston, and New Orleans-Alabama. Western production areas include Los Angeles and San Francisco.

Transportation routes between manufacturer and consumer spread throughout North America. Cotton plantations in the south require large amounts of DDT, whereas in the Midwest many tons of toxaphene and trifluralin are required for soybean crops. Hazardous materials pass through large cities, small rural communities, and the open countryside continually 24 hours a day. There is a need for various municipalities to be able to assess the hazards that exist in any community with regard to exposure to hazardous materials.

Our system defines a population and environmental rating scheme and a maximum disaster potential index. Rating values account for the highway, rail, marine, and pipeline transport of hazardous substances plus the industrial and storage handling of these substances.

The population hazard rating (HP) is given by:

$$HP = HP_H + HP_R + HP_m + HP_{IND} + HP_{ST}$$

where HP_H = population rating associated with the highway transportation of hazardous materials, and the subletters H, R, m, IND, and ST stand for highway, railway, marine, industry, and storage, respectively.

$$= (RPK)_{Route\ 1} + (RPK)_{Route\ 2} + \cdots + (RPK)_{Route\ n}$$

$$= \sum_1^n R_i P_i K_i$$

where n = number of routes

R = a route factor

$$= d\ C_H Z_H$$

where d = average traffic density in vehicles per hour (see Section 1 below)

C_H = fraction of vehicles carrying hazardous materials (see Section 2)

Z_H = 0.5 for divided highways

= 1.0 for dual highways

= 2.0 for all other highways

P = population density factor summation along a given route

$$= \sum_1^n m_i D_i$$

where n = number of population classifications

m = distance in route kilometers through a population density region of factor D (see Section 3)

K = a material classification given by:

$$= Q \sum_1^n A_i f_i$$

where Q = tonnage factor equal to unity (see Section 5)

A = a numerical population hazard index for substances of a given classification (see Section 4)

n = the number of classifications

f is a frequency fraction given by f = total tank trucks carrying material of a given classification per 100 tank trucks

$$HP_R = \sum_1^n R_i P_i K_i$$

where $R = NC_T Z_R$

N = number of trains per hour (daily average)

C_T = fraction of freight trains per total trains

Z_R = route coefficient

= 10 for a nonstop line through assessment area

= 5 for a marshalling yard

= 1 for a "quiet" branch line

P is determined as described previously

$$K = Q \sum_1^n A_i f_i$$

where Q = 50 (see Section 5)

A (see Section 4)

f = fraction of rail tank cars carrying material of a given classification per total cars of hazardous material

$$HP_m = \sum_1^n R_i P_i K_i$$

where $R = NC_m Z_m$

N = number of ships per hour (average of 24 hr)

C_m = total tankers and chemical barges

Z_m = unity for wide, slow-moving rivers

= 5 for areas having narrow bridges, etc.

= 10 for areas with narrow bridges, etc., and fast-moving currents

f = fraction of material of a given classification per

total vessels carrying hazardous products

$Q = 50$ (see Section 5)

The index for industry (HP_{IND}) is given by:

$$HP_{IND} = \sum_{1}^{n} (\Sigma m_i D_i)(\Sigma A_i Q_i)$$

where mD = kilometers of perimeter of an industrial complex at a population density factor of D (see Section 3)

AQ = a material classification where

A = the population hazard index for the material (see Section 4)

Q = tonnage factor (see Section 5)

HP_{ST} refers to warehouses and storage units that are not included within the rated perimeters of industry.

$$HP_{ST} = \sum_{1}^{n} D_i(\Sigma A_i Q_i)$$

A maximum disaster potential (DP) for the assessment area was formulated using the data on chlorine provided by Simmons [1] and the U.S. Department of Transportation [2]. Assuming that the observed area of a large spill of chlorine is 74.3 m^2 and that a 16-km/hr wind is blowing, the area of danger to the population will exhibit the following dimensions:

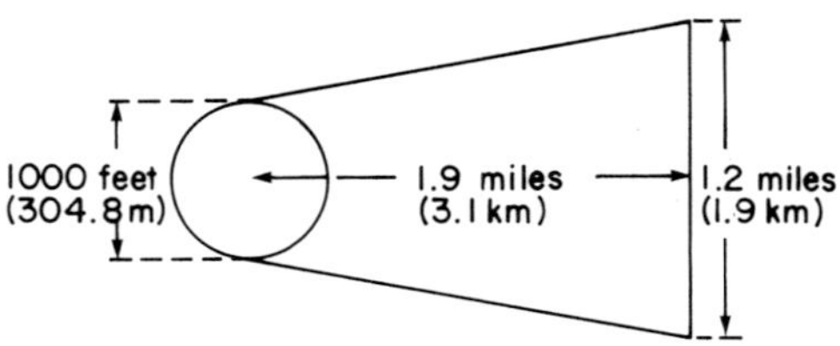

The area represented within the perimeter of the above diagram is approximately 6.5 km^2. The DP for the assessed area is postulated to be

$$DP = 0.5\ (6.48)(D_{max})$$

where D_{max} is the maximum population density in persons per km^2 along a transportation route or adjacent to industry.

When determining DP, select a combination of the highest population density with a highest value of A. The coefficient 0.5 assumes that one-half the people would be evacuated from the affected area before becoming disabled or expiring.

The environmental hazard index (HE) for the assessment area is given by:

$$HE = HE_H + HE_R + HE_m + HE_{IND} + HE_{ST} + HE_p$$

where p = pipeline.

Indices HE_H, HE_R, and HE_m are given by:

$$HE = \sum_1^n R_i m_i (T_1Q_1f_1 + T_2Q_2f_2 + \cdots + T_nQ_nf_n)_i$$

where R = route factor (described previously)

m = kilometers of route along or over navigable water or near sewers and ditches close to navigable water

T = ecological rating factor for a given classification of hazardous material with respect to water contamination (see Section 4)

Q* = tonnage factor (see Section 5)

f = fractional factor (see above)

n = number of routes

The remaining environmental factors are given by:

$$HE_{IND} = \sum_1^n m_i (T_1Q_1 + T_2Q_2 + \cdots + T_nQ_n)_i$$

$$HE_{ST} = \sum_1^n m_i T_i Q_i$$

---

*Q = 1 for HE_H, HE_{IND}, HE_{ST}, and HE_p

Q = 50 for HE_R, HE_m

$$HE_p = a \sum_{1}^{n} m_i T_i Q_i bc$$

where m = perimeter in km bordering on a navigable waterway or a creek or sewer leading to a navigable waterway. For pipelines, m includes under-river crossings

a = geological factor accounting for the earthquake potential of the area being assessed (see Section 6)

b accounts for surface or buried pipelines (see Section 6)

c accounts for the age of the pipeline (see Section 6)

n = number of industries, storage depots, and pipelines, respectively

The state of preparedness of the assessed community can be judged by the number of affirmative replies to the 14 statements in Section 7. Steps should be taken to examine the reasons for negative replies.

A calculation guide for the various indices in the rating system is presented in Section 8.

To test the effectiveness of the rating system, a small industrial city, similar to Sarnia, Ontario, was assessed. The parameters in Section 9 approximate the parameters for Sarnia. The methods for tabulating and calculating the necessary data are outlined in Section 9 and provide the reader with a model for the application of the rating system to other communities.

Reviewing Section 9, our model city showed a community population hazard index of approximately 19,000. Rail, marine, and industrial indices were highest. The maximum disaster potential was 6250, which is indeed ominous when considering the number of people involved. The environmental hazard index was about 11,100 of which HE_m accounted for 10,800. This high marine value was caused by the existence of a narrow, fast-current, highly traveled ship channel in the assessment area. Obviously this is an area that must be studied to minimize the possibility of a future mishap.

Table 1 indicates the magnitudes of population and environmental indices that might be expected after assessing various types of municipalities.

Table 1

Typical Hazard Magnitudes Expressed by the Community Rating Model[a]

Type of Community	HP[b]	DP[c]	HE[d]
Rural crossroads on a main highway	0.1	2	0.3
Village located by a main highway and rail line	2.5	3	3
Town	10	10	5
Small industrial city	150	100	50
Large industrial city	500	100	200

[a]All values × 10^2.

[b]HP = population hazard rating.

[c]DP = disaster potential.

[d]HE = environmental hazard index.

Section 1 AVERAGE TRAFFIC DENSITY

Highway traffic counts for the route under consideration should be carried out in 15-min intervals on Monday, Thursday, and Saturday at the times listed below. From these measurements and average traffic, density can be assumed.

Daily Times for Measurements

2:00- 2:15 a.m.
4:00- 4:15 a.m.
6:00- 6:14 a.m.
7:00- 7:15 a.m.
8:00- 8:15 a.m.
10:00-10:15 a.m.
12:15-12:30 p.m.
2:00- 2:15 p.m.
4:00- 4:15 p.m.

Daily Times for Measurements (continued)

5:00- 5:15 p.m.
5:30- 6:45 p.m.
8:00- 8:15 p.m.
10:00-10:15 p.m.
12:00-12:15 a.m.

Data on train movements can be obtained from the yardmaster of the area railway company. Figures on cargo shipments may also be available.

The harbor master of the assessment area will be helpful when establishing marine parameters.

Section 2 FRACTION OF VEHICLES CARRYING HAZARDOUS MATERIALS

When counting highway vehicles, record the number of tank trucks. Assume that tankers are the only carriers of hazardous cargos. Keep a record of individual tankers and their cargos.

Assume that every freight train hauls in excess of 500 metric tons of hazardous materials; this defines a tonnage factor of 50 for trains.

For marine vessels, consider as carriers of hazardous materials tankers, carriers of compressed gases, and barges containing liquid cargos. Assign a tonnage factor of 50.

Section 3 POPULATION DENSITY FACTOR

Population Density Range, Persons per km^2	Population Density Factor D
0 to 20	0.1
20 to 40	0.2
40 to 100	0.5
100 to 200	1.0
200 to 400	2.5

Population Density Range, Persons per km^2	Population Density Factor D
400 to 1000	5
1000 to 2000	10
2000 and above	25

Section 4 CLASSIFICATION OF HAZARDOUS MATERIALS [3]

A = numerical population hazard index

T = ecological rating factor

Population Hazard Classifications
(airborne substances)

Group Number	Type of Hazardous Material	Assigned Value for Factor A
0P	Nonvolatile, nonflammable, nontoxic solids and liquids	0
1P	Flammable, nontoxic solids and liquids with noncorrosive or toxic vapors	1
2P	Flammable substances that are thermally stable but quite volatile and release relatively nontoxic vapors in their natural or burning state	2
3P	Flammable, thermally unstable or corrosive substances	5
4P	Substances that release toxic vapors either burning or otherwise	10

Environment Hazard
Classification (for water)

Group Number	Type of Hazardous Material	Ecological Rating Factor T	TLm Concentration Aquatic Toxicity, mg/liter
0E	Minimal hazard, low toxicity and low persistence	0	Above 1000
1E	Slight hazard, low toxicity, high persistence to moderate toxicity, low persistence	2	100-1000
2E	Moderate hazard, moderate toxicity to high persistence	5	10-100
3E	Highly hazardous materials, moderate to high toxicity, high persistence	8	1-10
4E	Extreme hazards, high toxicity, very high persistence	10	Below 1

	Hazard Factor*	
Hazardous Material	Population Hazard Index for Airborne Substances, A	Ecological Rating Factor for Material in Water, T
Acetone	2	0
Acetone cyanohydrin	1	8
Acetonitrile	3	2
Acetylene	5	2
Acrolein	10	8
Acrylic acid	1	6
Acrylonitrile	2	5
Aluminum chloride	0	10
Ammonia	2 (nonflammable)	5
Ammonium nitrate	0	2
Aniline	1	5
Antimony compounds	0	5
Arsenic compounds	0	9

Hazardous Material	Hazard Factor*	
	Population Hazard Index for Airborne Substances, A	Ecological Rating Factor for Material in Water, T
Barium compounds (soluble)	0	4
Benzene	1	8
Benzene sulfonic acid	1	2
Benzoic acid	1	2
Beryllium compounds	0	5
Boron compounds	0	0
Bromine	2	5
Butadiene	2	5
Butane	2	0
Butyl mercaptan	5	10
Cadmium compounds	0	8
Carbon disulfide	2	5
Carbon tetrachloride	1	2
Caustic (NaOH and KOH)	0	5
Crude oil	1	2
Chlorine	10	10
Chlorobenzene	1	5
Chromium compounds	0	8
Copper compounds	0	10
Cyclohexane	1	6
Chlorinated hydrocarbons	1	10
Cyanides (inorganic other than HCN)	5 (acids present)	10
Detergents	0	5
Dimethyl ether	2	0
Ethers (general, low mol wt)	2	0
Ethyl acetate	2	1
Ethyl alcohol	2	0
Ethylamine	2	5
Ethylbenzene	1	5

Hazardous Material	Hazard Factor*	
	Population Hazard Index for Airborne Substances, A	Ecological Rating Factor for Material in Water, T
Ethylchloride	2	0
Ethylene	2	5
Ethylenediamine	1	5
Ethylene glycol	1	0
Ethylene glycol monoether	1	2
Ethylene oxide	5	2
Ethylinimine	10	7
Ethyl ether	2	0
Ferric salts (soluble)		10
Fluorine	10	10
Formaldehyde	2	5
Fuel oils (light, diesel)	1	0
Glycerin	1	0
Gasoline	2	1
Herbicides (all)	3	10
Heptane	2	0
Hexane	2	0
Hydrochloric acid	0	8
Hydrofluoric acid	3	5
Hydrogen (liquid)	2	0
Hydrogen chloride	8	8
Hydrogen cyanide	10	10
Hydrogen peroxide	5	5
Hydrogen sulfide	10	10
Hypochlorite	0	10
Hydrocarbon fuels (general)	2	0
Hydrogen fluoride	10	2
Insecticides (all)	3	10
Isobutene	2	0

Hazardous Material	Hazard Factor*	
	Population Hazard Index for Airborne Substances, A	Ecological Rating Factor for Material in Water, T
Isooctane	1	0
Isoprene	2	5
Iron compounds (soluble)		10
Industrial wastes (containing metals, phenols, etc.)	0	10
Lead compounds	0	8
Latex, rubber	0	8
Mercury compounds (inorganic)	0	10
Methyl alcohol	1	2
Methylchloride	2	0
Methyl mercaptan	3	10
Mercaptans	3	10
Methane (liquid)	2	0
Methyl ethyl ether	2	0
Methylamines	5	1
Nickel compounds	0	8
Nitric acid	5	8
Nitrobenzene	1	5
Nitrogen tetroxide	5	0
Oxygen (liquid)	2	0
Propane/LPG	2	0
Pentane	2	0
Perchloric acid	5	8
Petroleum ether	2	0
Phenol	1	10
Phosphine	10	8
Phosphoric acid	0	7
Phosphorus	5	10
Phosphorus trichloride	5	8

Hazardous Material	Hazard Factor*	
	Population Hazard Index for Airborne Substances, A	Ecological Rating Factor for Material in Water, T
Propylene	2	0
Propyl alcohols	1	2
Propylene oxide	2	0
Pesticides (all)	2	10
Radioactive materials	10	10
Selenium (colloidal)	10	8
Silver compounds	0	10
Styrene	2	5
Sulfur (slate, pellets)	0	0
Sulfuric acid	0	8
Triethylamine	1	5
Tetraethyl lead	10	10
Toluene	1	5
Vinyl chloride	2	5

*Postulated by the authors based on the literature. These values are subject to future modifications.

Section 5 TONNAGE FACTORS

Total Vehicle Capacity, metric tons	Assigned Tonnage Factor, Q
0-10	1
10-100	10
100-500	25
500+	50

Section 6 PIPELINE FACTORS

Location	Assigned Geological Factor, *a*
Rocky Mountains	10
Prairies, western Ontario	1
Eastern Canada	3

Buried pipeline	Assign value *b* = 1
Pipeline above ground	Assign value *b* = 10
Pipelines >10 yr old	Assign value *c* = 2
Pipelines <10 yr old	Assign value *c* = 1

Section 7 ASSESSMENT OF PREPAREDNESS
(Circle appropriate answers)

1. An emergency plan exists for or includes the assessment area and involves municipal and community governments and industry. YES NO
2. If an emergency plan exists for or includes the assessment area, this plan has been formulated into a written document defining responsibilities and authority. YES NO
3. All industries handling hazardous materials in the assessment area have individual, in-plant emergency plans which are compatible with a general community plan. YES NO
4. The emergency plan for the assessment area has a clearly defined procedure for the sharing of hardware and expertise with external resources. YES NO
5. The emergency plan for the assessment area is periodically tested by the simulation of emergencies and is upgraded. YES NO

6. The emergency plan has undergone actual testing in a real emergency and this was followed by a detailed analysis on the functionability of the plan. YES NO
7. Truck routes have been specified in the assessment area. YES NO
8. Within the past year there have been at least two speeding convictions involving tank trucks. YES NO
9. Within the past year there have been at least two convictions involving truckers not having followed truck routes. YES NO
10. Board of Transport speed regulations for trains are strictly enforced in the assessment area. Train speeds are constantly monitored and violators prosecuted and fined. YES NO
11. Rails are aligned and leveled in the assessment area at least once a year, using special rail-testing devices designed for this purpose. YES NO
12. A hardware inventory including heavy equipment, adsorbants, containment devices, etc., has been established by a planning committee for emergencies involving the assessment area. YES NO
13. Sewer shut-off points exist for the retainment of hazardous materials that may leak into sanitary and storm sewers. YES NO
14. A spot-check of three tank truck drivers has revealed that each knew what action to take in event of the spilling of hazardous materials. YES NO

In calculating the hazard rating of a community, many formulas and terms are used. For convenience, all of the terms are defined below, followed by the assessment procedure.

A Population hazard index for a given substance (Section 4)

a Geological factor accounting for earthquake potential (Section 6)

b Factor accounting for surface or buried pipelines (Section 6)

C	Fraction of vehicles carrying hazardous materials
C_H	C value for highway transportation (Section 2)
C_m	Total tankers and chemical barges
C_T	Fraction of freight trains
c	Factor accounting for the age of a pipeline (Section 6)
D	Population density factor used in the calculation of P (Section 3)
D_{max}	Maximum population density in persons per km^2
d	Average traffic density in vehicles per hour (Section 1)
DP	Disaster potential
E	Environmental
f	Frequency fraction
HE	Environmental hazard index
HE_H	HE value for highway transportation
HE_{IND}	HE value for industry
HE_m	HE value for marine transportation
HE_P	HE value for pipelines
HE_R	HE value for rail transportation
HE_{ST}	HE value for the storage of hazardous materials
K	Classification index for a substance
m	distance in route km
n	Number of classifications, groups, routes, etc.
N	Average number of trains or ships per hour
P	Population density factor
Q	Tonnage factor (Section 5); 1 for HE_H, HE_{IND}, HE_P, HE_{ST}
R	Route factor
R_H	R value for highway transportation
R_m	R value for marine transportation
R_R	R value for rail transportation
T	Ecological rating factor for a given classification of hazardous materials (Section 4)
Z	Route subfactor used in the calculation of R
Z_H	Highway factor 0.5 for divided highways 1.0 for dual highways 2.0 for all other highways

Z_m	Marine route factor 1 for wide, slow-moving rivers 5 for areas having narrow bridges or some other form of obstruction on a slow-moving river 10 for areas having narrow bridges or some other form of obstruction on a fast-moving river
Z_R	Rail route factor 1 for a "quiet" branch line 5 for a marshalling yard 10 for a nonstop line through the assessment area

SUBSCRIPTS

H	highway
IND	industry
m	marine
P	pipeline
R	rail
ST	storage

The following charts have been prepared for those interested in making a hazard analysis. All that is required is care in filling in the numbers properly and summing them to obtain the proper rating.

Section 8 COMMUNITY HAZARD RATING SYSTEM

Calculation Guide

Community assessed ____________________

State or province, etc. ____________________
____
____

Population ____________________

Type of community ____________________
____
____
____

Assessed by ____________________

Title ____________________
____
____

Date of assessment ____________________

Calculation of HP_H, HP_R, HP_m
(circle which)

Route Number:	1	2	3	4	5	6	7
d, N							
C, C_T, C_m							
Z, Z_R, Z_m							
R_H, R_R, R_m							
mD for D = 0.1							
0.2							
0.5							
1.0							
2.5							
5.0							
10							
25							
$P = \Sigma$ mD							
QfA for A = 1							
2							
5							
10							
$K = Q \Sigma Af$							
RPK							

$HP_{\square} = \Sigma$ RPK = ________

Note: See pages 144-148 or 158-160 for definition of terms.

Calculation of HP_{IND}
(Hazard Index — Industry)

Industry Number:	1	2	3	4	5	6	7
m for D = 0.1							
0.2							
0.5							
1.0							
2.5							
5.0							
10							
25							
Σ mD							
Q for A = 1							
2							
5							
10							
Σ AQ							
(Σ mD)(Σ AQ)							

$HP_{IND} = \Sigma\ (\Sigma md)(\Sigma AQ) =$ ________

Calculation of HP_{ST}
(Hazard Index — Storage)

Storage Unit Number:	1	2	3	4	5	6	7
D							
Q for A = 1 2 5 10 Σ AQ							
D(Σ AQ)							

$HP_{ST} = \Sigma\, D(\Sigma AQ) =$ ________

$HP = HP_H + HP_R + HP_m + HP_{IND} + HP_{ST}$

= ______ + ______ + ______ + ______ + ______

= __________ = community population hazard index

Calculation of DP
(Maximum Disaster Potential)

$DP = 3.24\ D_{max} = 3.24$ ________ = ________

Note: See page 146 for definition of terms.

Calculation of HE_H, HE_R, HE_m
(circle which)

Route Number:	1	2	3	4	5	6	7
R m							
QfT for T = 2 5 8 10 Σ TQf							
Rm(Σ TQf)							

$HE_{\square} = \Sigma\ Rm(\Sigma\ TQf) =$ ________

Note: See pages 147, 148 or 158-160 for definition of terms.

Calculation of HE_{IND}

Industry Number:	1	2	3	4	5	6	7
m							
Q for T = 2 5 8 10 m(Σ QT)							

$HE_{IND} = \Sigma\ m(\Sigma QT) =$ ________

Calculation of HE_{ST}

Storage Unit No.:	1	2	3	4	5	6	7
m							
QT for T = 2 5 8 10 Σ QT							
m(Σ QT)							

$$HE_{ST} = \Sigma\, m(\Sigma\, QT) = \_\_\_\_\_\_$$

Calculation of HE_p

Pipeline Number:	1	2	3	4	5	6	7
Area <u>a</u> Value							
m							
QT for T = 2 5 8 10 Σ QT							
b c							
mbc Σ QT							

$$HE_p = a\, \Sigma\, mbc(\Sigma\, QT) = \_\_\_\_\_\_$$

$$HE = HE_H + HE_R + HE_m + HE_{IND} + HE_{ST} + HE_p$$

$$= \_\_\_\_ + \_\_\_\_ + \_\_\_\_ + \_\_\_\_ + \_\_\_\_ + \_\_\_\_$$

$$= \_\_\_\_\_\_ = \text{environmental hazard index}$$

Section 9 COMMUNITY HAZARD RATING SYSTEM II

Example Calculation for a Highly Industrialized Area with a Major Waterway and Close to a Lake

Community Assessed	City X
Province, etc.	—
Population	60,000
Type of Community	Industrial
Assessed by	The Authors
Title	The University of Western Ontario Study Project
Date of Assessment	January 1976

Traffic Density Assessment for Three Truck Routes

Route Number:	1			2			3		
Day:	Mon.	Thu.	Sat.	Mon.	Thu.	Sat.	Mon.	Thu.	Sat.
Time									
2:00- 2:15 a.m.	6	3	5	2	1	5	2	3	1
4:00- 4:15 a.m.	2	3	2	3	1	0	3	6	0
6:00- 6:15 a.m.	10	7	8	7	6	6	5	2	2
7:00- 7:15 a.m.	15	14	15	10	8	8	10	8	5
8:00- 8:15 a.m.	25	30	20	30	35	5	15	10	7
10:00-10:15 a.m.	18	17	16	25	22	30	18	15	20
12:15-12:30 p.m.	20	22	21	35	40	25	10	8	15
2:00- 2:15 p.m.	17	16	15	20	22	30	12	10	14
4:00- 4:15 p.m.	20	22	18	22	25	26	15	12	16
5:00- 5:15 p.m.	35	30	20	30	35	25	20	18	15
5:30- 5:45 p.m.	30	32	20	25	20	26	15	12	8
8:00- 8:15 p.m.	10	8	8	20	18	22	8	6	7
10:00-10:15 p.m.	8	7	6	10	8	15	9	7	7
12:00-12:15 a.m.	5	4	5	2	3	1	4	3	3
Average Vehicles									
15 min	16	15	13	17	17	16	10	9	9
1 hr	64	60	52	68	68	64	40	36	36
Average/hr (d)	59			67			37		
Route Factor Z_H	0.5			1.0			2.0		

Vehicles Carrying Hazardous Cargos
(All Tank Trucks)

Route Number:	1			2			3		
Day:	Mon.	Thu.	Sat.	Mon.	Thu.	Sat.	Mon.	Thu.	Sat.
Time									
2:00- 2:15 a.m.	2	3	2	0	1	0	1	2	0
4:00- 4:15 a.m.	1	2	2	0	0	0	1	2	0
6:00- 6:15 a.m.	4	3	2	1	0	0	0	1	0
7:00- 7:15 a.m.	6	5	5	2	1	2	1	1	1
8:00- 8:15 a.m.	2	8	2	3	2	4	3	2	3
10:00-10:15 a.m.	3	4	1	2	3	3	2	3	3
12:15-12:30 p.m.	2	4	3	3	3	3	1	0	2
2:00- 2:15 p.m.	1	4	3	4	3	4	2	3	2
4:00- 4:15 p.m.	3	0	4	3	0	1	3	2	2
5:00- 5:15 p.m.	4	3	5	2	2	1	2	4	1
5:30- 5:45 p.m.	4	3	2	2	1	2	0	2	2
8:00- 8:15 p.m.	5	3	1	1	0	0	0	1	0
10:00-10:15 p.m.	2	1	0	0	1	0	2	1	0
12:00-12:15 a.m.	1	2	0	0	0	1	2	0	0
Average Vehicles 15 min	3	3	2	1.6	1.2	1.5	1.4	1.7	1.1
Average/hr	11			5.7			5.6		
C	C.187			0.085			0.152		
$R = dC\ Z_H$	5.5			5.7			11.3		

Train Data

	Route: Main Line	Route: Alternate
Z_R	5	1
Average total trains/day	15	1
Average freight trains/day	9	1
Average trains/hr (N)	0.62	0.04
C	0.6	1
R ($NC\ Z_R$)	1.9	0.04

Ship Data

	Route: River
Z_m	10
Average total ships/day	48
Average tankers, barges, etc.	5
C_m	0.083
Average ships/hr (N)	2
R_m (NC_mZ_m)	1.7

Cargo Data

Highway (Tankers)

Material	A	T	Route Number 1 (Percent)	Route Number 2 (Percent)	Route Number 3 (Percent)
Gasoline	2	0	20	25	10
Fuel oil	1	0	15	30	5
Asphalt	0	0	2	10	0
Caustic	0	5	15	7	20
Ammonia solutions	0	5	15	0	25

Material	A	T	Route Number 1	2	3
			Percent		
Acids (general)	0	8	15	15	15
Industrial waste	0	10	0	0	5
Ethylenimine	10	8	2	0	0
Chlorine	10	10	3	0	0
Oxygen	2	0	1	0	3
Propane	2	0	2	0	5
Benzene	1	8	5	10	10
Chemicals (general)	5	5	5	0	0
Ammonia	2	5	0	3	2
Total			100	100	100

Railway (Tank Cars)

Material	A	T	Route Number 1	2
			Percent	
Chlorine	10	10	13.8	0
Caustic	0	5	20	30
Benzene	1	8	5	0
Ammonia (anhydrous)	2	5	15	40
Sulfuric acid	0	8	10	0
Hydrochloric acid	0	8	4	0
Methyl alcohol	1	2	2	0
Ammonium nitrate solutions	0		5	15
Phenol	1	10	5	0
Styrene	2	5	4	0
Rubber latex	0	8	5	0

Table (continued)

Material	A	T	Route Number 1	2
			Percent	
Ethylene oxide	5	2	3	0
Ethylene glycol	1	0	5	10
Fluorine	10	10	0.1	0
Nitric acid	5	8	1.0	0
Tetraethyl lead	10	10	0.1	0
Vinyl chloride	2	5	2.0	5
Total			100	100

Marine

Material	Percent
Gasoline and fuel oil	80
Chlorine	5
Propane/LPG	5
Sulfuric acid	5
Ammonia (anhydrous)	3
Caustic	2
Total	100

Industry

1. Petrochemical complex and petroleum refining
   perimeter = 5 kilometers
   water perimeter (m) = 2 kilometers
2. Chemical complex (nitric acid and ammonia)
   perimeter = 4 kilometers
   water perimeter (m) = 0 kilometers

Storage

1. Two fuel oil depots

	No. 1	No. 2
Population factor D	10	1
Tonnage	— 500+ —	

   No water contamination hazard

2. Chlorine storage

   Population factor, D = 0.5

   Tonnage = 500+

   No water contamination hazard

Pipelines

1. Natural gas, buried, no water nearby, pipeline 12 years old.
2. Crude oil, buried, passes 1 km under river, pipeline 14 years old.

Example Calculation of $\textcircled{HP_H}$, HP_R, HP_m

Route Number:	1	2	3	4	5
$\textcircled{d}$, N	59	67	37		
$\textcircled{C_H}$, C_T, C_m	0.187	0.085	0.152		
$\textcircled{Z_H}$, Z_R, Z_m	0.5	1.0	2.0		
$\textcircled{R_H}$, R_R, R_m	5.5	5.7	11.3		
mD for D = 0.1					
0.2					
0.5			6 x 0.5		
1.0					
2.5					
5.0	2 x 5	1 x 5			
10	2 x 10	1 x 10			
25		2 x 25			
$P = \Sigma$ mD	30	65	3		
fA for A = 1	0.2 x 1	0.4 x 1	0.15 x 1		
2	0.23 x 2	0.28 x 2	0.20 x 2		
5	0.05 x 5				
(Q = 1) 10	0.05 x 10				
$K = Q \Sigma Af$	1.4	0.96	0.55		
$R_H PK$	231	356	18.7		

$$HP_{\boxed{H}} = \Sigma R_H PK = \underline{606}$$

Example Calculation of HP_H, HP_R (circled), HP_m

Route Number:	1	2	3	4	5
d, N (circled)	0.62	0.04			
C_H (circled), C_T, C_m	0.6	1			
Z_H, Z_R (circled), Z_m	5	1			
R_H, R_R (circled), R_m	1.9	0.04			
mD for D = 0.1					
0.2					
0.5	6 x 0.5				
1.0					
2.5	3 x 2.5	4 x 2.5			
5.0					
10	1 x 10				
25					
$P = \Sigma$ mD	20.5	10			
fA for A = 1	0.17 x 1	0.10 x 1			
2	0.21 x 2	0.45 x 2			
5	0.04 x 5				
(Q = 50) 10	0.14 x 10				
$K = Q\ \Sigma$ Af	110	50			
$R_R PK$	4300	20			

$HP_{\boxed{R}} = \Sigma\ R_R PK =$ 4320

Example Calculation of HP_H, HP_R, ⒽHP_m (circled)

Route Number:	1	2	3	4	5
d, (N) (circled)	2				
C_H, C_T, (C_m) (circled)	0.083				
Z_H, Z_R, (Z_m) (circled)	10				
R_H, R_R, (R_m) (circled)	1.7				
mD for D = 0.1					
0.2					
0.5					
1.0	4 x 1				
2.5					
5.0	2 x 5				
10	1 x 10				
25					
$P = \Sigma mD$	24				
fA for A = 1					
2	0.88 x 2				
5					
(Q = 50) 10	0.05 x 10				
$K = Q \Sigma Af$	113				
$R_m PK$	4600				

$HP_{\boxed{m}} = \Sigma R_m PK =$ 4600

Example Calculation of HP_{IND}

(Hazard Index — Industry)

Industry Number:	1	2	3	4	5
m for D = 0.1	3 x 0.1				
0.2					
0.5					
1.0	2 x 1				
2.5					
5.0		4 x 5			
10					
25					
Σ mD	2.3	20			
Q for A = 1	50				
2	50	50			
5	50	50			
10	50				
Σ AQ	900	350			
(Σ mD)(Σ AQ)	2070	7000			

$$HP_{IND} = \Sigma\ (\Sigma\ mD)(\Sigma\ AQ) = \underline{\ 9070\ }$$

Example Calculation of HP_{ST}

(Hazard Index — Storage)

Storage Unit Number:	1	2	3	4	5
D	10	1	0.5		
Q for A = 1	1 x 50	1 x 50			
2					
5					
10			10 x 50		
Σ AQ	50	50	250		
D(Σ AQ)	500	50	250		

$$HP_{ST} = \Sigma\, D(\Sigma\, AQ) = \underline{\quad 800 \quad}$$

$$HP = HP_H + HP_R + HP_m + HP_{IND} + HP_{ST}$$

$$= \underline{\ 606\ } + \underline{\ 4320\ } + \underline{\ 4600\ } + \underline{\ 9070\ } + \underline{\ 800\ }$$

$$= \underline{\ 19396\ } = \text{Community Population Hazard Index}$$

Example Calculation of DP

(Maximum Disaster Potential)

$$DP = 3.24\ D_{max} = 3.24\ \underline{\ 1929\ } = \underline{\ 6250\ }$$

Example Calculation of (circled) HE_H, HE_R, HE_m

Route Number:	1	2	3	4	5
R_H	5.5	5.7	11.3		
m	1	0.5	0		
fT for T = 2					
5	0.35 x 5	0.47 x 5			
8	0.22 x 8	0.25 x 8			
(Q = 1) 10	0.03 x 10	0.05 x 10			
Σ TQf	3.81	4.85			
$R_H m(\Sigma$ TQf)	21	14	0		

$HE_{\boxed{H}} = \Sigma R_H m(\Sigma TQf) =$ 35

Example Calculation of HE_H, HE_R (circled), HE_m

Route Number:	1	2	3	4	5
R_R	1.9	0.04			
m	0.5	0			
fT for T = 2	0.05 x 2				
5	0.41 x 5				
8	0.25 x 8				
(Q = 50) 10	0.19 x 10				
Σ TQf	303				
$R_R m(\Sigma\ TQf)$	287				

$$HE_{\boxed{R}} = \Sigma\ R_R m(\Sigma\ TQf) = \underline{\quad 287 \quad}$$

Example Calculation of HE_H, HE_R, HE_m (circled)

Route Number:	1	2	3	4	5
R_m	1.7				
m	7				
fT for T = $\not{2}$	0.8 x 1*				
5	0.05 x 5				
8	0.05 x 8				
(Q = 50) 10	0.05 x 10				
Σ TQf	975				

*Let gasoline and fuel oils have an A value of 1.

$$HE_m = \Sigma\ R_m m(\Sigma\ TQf) = \underline{\quad 10800 \quad}$$

Example Calculation of HE_{IND}

(Hazard Index — Industry)

Industry Number:	1	2	3	4	5
m	2	0			
QT for T* = 2	0.5 x 2				
5	0.05 x 5				
8	0.05 x 8				
(Q = 1) 10	0.05 x 10				
Σ QT	2.15				
m(Σ QT)	4.3				

*Assumed values.

$HE_{IND} = \Sigma\, m(\Sigma\, QT) =$ 4.3

Example Calculation of HE_{ST}

(Hazard Index — Storage)

Storage Unit Number:	1	2	3	4	5
m					
QT for T = 2					
5					
8					
10					
Σ QT					
m(Σ QT)					

$HE_{ST} = \Sigma\, m(\Sigma\, QT) =$ 0

Example Calculation of HE_P

Pipeline Number:	1	2	3	4	5
Area a value	1	1	1		
m	0	1	1		
QT for T = 2		1 x 2	1 x 2		
5					
8					
(Q = 1) 10					
Σ QT		2	2		
b		1	1		
c		2	2		
mbc Σ QT		4	4		

$$HE_P = a\ \Sigma\ mbc(\Sigma\ QT) = \underline{\ 8\ }$$

$$HE = HE_H + HE_R + HE_m + HE_{IND} + HE_{ST} + HE_P$$

$$= \underline{35} + \underline{287} + \underline{10800} + \underline{4.3} + \underline{0} + \underline{8}$$

$$= \underline{11134}$$ = Environmental Hazard Index

REFERENCES

1. J. A. Simmons et al., Risk Assessment of Large Spills of Toxic Materials, Proc. 1974 National Conf. Control of Hazardous Material Spills, San Francisco, August 25-28, 1974, pp. 166-175.

2. U.S. Dept. of Transportation, Emergency Services Guide for Selected Hazardous Material Spills, Fire, Evacuation Area, Office of the Secretary of Transportation, Washington, 1973.

3. R. W. Hahn and P. A. Jensen, The Importance of Waterway Dilution Capacity in Hazardous Material Spills and a Method to Include This Parameter in a Risk Decision Framework, Proc. 1974 National Conf. on Control of Hazardous Material Spills, San Francisco, August 25-28, 1974, pp. 349-352.

APPENDICES

Appendix A

RECENT SPILL STATISTICS

Table A-1

Hazardous Materials Spilled[a] in Ontario, January 1, 1971 to July 31, 1973

Material	Quantity
Spilled from trucks	
Edible oil	15.9 m^3
Calcium phosphate	18,144 kg
Ammonium nitrate solution	23,133 kg
Phosphoric acid	0.45 m^3
Hydrochloric acid	3629 kg
Solution of NH_3, NaCl, KCl, trace metals	0.45 m^3
Diethanolimide	11.4 m^3
Ferric chloride	15.4 m^3
Caustic solution	0.91 m^3
Alcohol	4.09 m^3
Copper sulfate	38 bags
Ammonia solution	9.1 m^3
Spilled from rail car	
Magnesium hydroxide	72.7 m^3
Copper concentrate	154.5 metric tons
Zinc concentrate	63.5 metric tons
Rape seed oil	13.6 m^3
Acetic anhydride	31.8 m^3
Storage/Transfer	
Spilled from valve	
Nitric acid	27.3 kg
Spilled from line	
Pickle liquor	27.3 m^3
Latex	30,618 kg

Table A-1 (continued)

Material	Quantity
Spilled from tank	
Brewery waste	2.3 m^3
Glycol	6.8 m^3
Methyl chloroform	1.14 m^3
Phosphoric acid	181.8 kg
Diisooctyl azelate	19.5 m^3
Polyvinyl chloride slurry	1.02 kg
Ammonia solution	6.8 m^3
Brine with 9 ppm Hg	2.3 m^3
Ethylene dichloride	4.55 m^3
Pickle acid	7.27 m^3
Brine with 5 ppm Hg	100.0 m^3
Calcium chloride solution	4.55 m^3
Caustic solution (5%)	4.55 m^3
Chromic acid	5.46 m^3
Vegetable oil	0.91 m^3
Turpolene solvent	1.23 m^3
Sulfuric acid	845.6 kg
Hydrochloric acid (3%)	22.7 m^3
Paint thinner	1.82 m^3
Hydrochloric acid	29.5 m^3
Spilled from pipeline	
Ammonia (anhydrous)	1.82 m^3

Source: Environment Canada, Spills of Oil and Hazardous Materials, Canadian Research and Development, People and Priorities, Rep. EPS-3EE-731, Environmental Emergency Branch, Ottawa, 1973.

[a]Petroleum spills are not included.

Table A-2

Spills of Hazardous Materials in Ontario, January 1, 1971 to July 31, 1972

Material	Total Spilled, %		
	Trucks	Rail	Storage/Transfer
Ammonia and ammonia solutions	23		8
Acidic solutions (inorganic)	15	17	36
Basic solutions (inorganic)	9	17	4
Organic liquids	23	17	32

Table A-2 (continued)

Material	Total Spilled, % Trucks	Rail	Storage/Transfer
Solids	15	49	
Other	15		20
Number of accidents	13	6	25

Source: Environment Canada, Spills of Oil and Hazardous Materials, Canadian Research and Development, People and Priorities, Rep. EPS-3EE-731, Environmental Emergency Branch, Ottawa, 1973.

LIST OF SPILLS OF HAZARDOUS MATERIALS (OTHER THAN OIL), JUNE 1, 1967 TO JUNE 1, 1970, U.S. FEDERAL WATER QUALITY ADMINISTRATION

Highly Toxic Substances

1. Acetone Cyanohydrin

   1/28/68, Dunreith, Ind.
   4.54 m^3 (1270 kg) from R.R. tank car (Penn Central R.R. and Norfolk and Western R.R.).
   Fish killed, cattle died after drinking water.
   Buck Creek, a tributary of Big Blue River.
   Cyanide liberated when chemical entered alkaline waters. HTH added to stream to neutralize cyanide. Residents of town evacuated. Penn Central R.R. reimbursed state for chemical costs.

2. Cyanide Compound and Hydrochloric Acid

   2/2/68, Lampson, Ala.
   Unknown quantity, two railroad cars, Southern R.R.
   Residents of town evacuated. No report on damage.
   Reported small escape of cyanide and acid, all retained by dike.

3. Vinyl Cyanide

   8/18/69, Cambridge, Pa.
   18.9 m^3, tank car turned over, American Cyanamid.
   Did not reach watercourse. No damage reported.
   Sodium carbonate used to neutralize material that did not vaporize.

4. Pesticides

    8/22/69, James River, Va.
    Large quantity, formulation plant, Miller Chemical Co.
    No damages reported.
    Plant flooded by rains of Hurricane Camille.

5. Pesticides

    8/25/69, Biloxi, Miss.
    Large quantities, storage warehouse, company unknown
    No damages reported.
    Material discharged into Mississippi River.
    Warehouse destroyed by Hurricane Camille.

6. Unknown Chemicals

    9/2/69, Escambia Bay, Fla.
    Large quantities (227 to 4536 kg), chemical plant, Escambia Chemical Co.
    Fish killed.

7. Chrome Wastes

    9/24/69, Hoosic River, Vt.
    Large quantities, tannery, company unknown.
    Fish killed, number unknown.

8. Pesticides

    10/3/69, Locust Fork of Black Warrior River, Birmingham, Ala.
    Quantity unknown, malathion suspected, company unknown.
    Estimated 750,000 fish killed.
    Low dissolved oxygen, algae killed, malathion found in water.

9. DDT

    Flint River, Albany, Ga.
    Unknown quantity, unknown company.
    Small fish kill.

<u>Toxic Gases</u>

1. Hydrogen Sulfide

    3/25/70, Jackson, Miss., Shell Oil Co.
    Wildcat oil well exploded and fire resulted.
    Fire consumed H_2S initially.

2. Chlorine

    11/18/68, Ohio River.
    Liquid chlorine barge.
    $5300 property damage.

<u>Corrosive Liquids</u>

1. Sulfuric Acid

    A. 6/9/69, Lewes, Del.
        4536 metric tons dumped into ocean.

B. 10/6/69, Black Creek, Purvis, Miss.
320-480 liters, Gulf Oil Refinery.
350,000 fish killed.
A dike will be constructed to control any such future events. Damages $20,650.

C. 11/13/69, Ocoee River, Polk County, Tenn.
17.8 m^3.
A dead river biologically, therefore no fish killed.
Lime used to neutralize acid.

D. 4/16/69, Lobdell, La.
One tank car of acid flushed into drains leading to the Mississippi River.
No further information.

E. 12/24/68, Houston, Texas.
406 metric tons from ship.
Total damage $13,025.
No information on cleanup.

F. 1/16/70, Mississippi River, barge.
$2000 damage.

G. 2/16/69, Atchafalaya River, barge.
$162,000 damage.

H. 1/20/69, Mississippi River, barge.
$4273 damage.

2. Phosphoric Acid

12/3/66, Mississippi River, barge.
1905 metric tons.
Total damage $68,000.
No information on cleanup.

3. Waste Acid

3/20/69, San Francisco, barge.
$10,000 damage.

<u>Other Materials</u>

1. Copper Refining Wastes

7/9/69, Morenci, Ariz.
1893 m^3, dike failure, Phelps Dodge Co.
50,000 fish killed in Chase Creek and San Francisco River.
Case pending.

2. Phenol and Acetone

5/3/70, Susquehanna River, Pa.
26.5 m^3, R.R. wreck, damages unknown.

3. Mixed Spill — Railway

1/1/68, Dunreith, Ind., Buck Creek
Acetone cyanohydrin — 158 m^3
Methyl methacrylate — 77.6 m^3

Ethylene oxide — 77.6 m^3
Vinyl chloride — 127 m^3
Heavy river contamination (see no. 1, Highly Toxic Substances).

4. Sulfur

   A. 8/13/68, Mississippi River, barge
   $355,000 damage.

   B. 7/24/68, Bayou Lafourche, La., liquid sulfur
   $3300 property damage.

   C. 12/27/68, Bayou Lafourche, La., liquid sulfur
   $2000 property damage.

Spills Involving Large Fish Kills
Other Than Those Mentioned Above

1. 6/10/67, Carbo, Va., alkaline fly ash slurry.
   Failure of a 162-ha diked reservoir.
   216,000 fish killed, Appalachian Power Co.

2. 6/17/69, Pensacola Bay, Fla.
   9072 kg undetermined material.
   9072 kg fish killed.

3. 9/23/69, Deer Creek, Ind.
   Anhydrous ammonia, Erney Fertilizer Plant.
   10,000 to 15,000 fish killed.

4. 10/2/69, Cahaba River, Centerville, Ala.
   Wood-treating chemicals, composition unknown.
   W. E. Belcher Lumber Co., 10,000 fish killed.

5. 5/26/70, Drakes Bay, Calif., Johnson Oyster Co.
   160 liters wood preservative.
   $10,000 worth of clams and crabs ruined.

6. 8/2/68, Cumberland, Tenn.
   Ethylene glycol and ethyl alcohol waste.
   Estimated 175,000 fish killed.

Source: U.S. House Document 92-70, Control of Hazardous Polluting Substances, A Report on Control of Hazardous Polluting Substances pursuant to Section 12(g) of the Federal Water Pollution Control Act as amended, U.S. Govt. Print. Off., Washington, 1971.

Appendix B

EMERGENCY DATA AND ACTION SYSTEMS

CANADA AND U.S. EMERGENCY DATA AND ACTION SYSTEMS, AND AUTHORITIES ON HAZARDOUS MATERIAL SPILLS
INFORMATION SHARED BETWEEN CANADA AND THE UNITED STATES

American Water Works Association Emergency Manual for Hazardous Materials Spills

The American Railroad Association

ADAPTS, Air Delivery Anti-Pollution Transfer System (U.S. Coast Guard)

The Bureau of Explosives

CAN-OLE, A National Science Library "on-line" Enquiry System located in Ottawa, Canada

CHLOREP, The Chlorine Emergency Plan (The Chlorine Institute)

CHEMCARD, initiated in 1964 by MCA. Cards are attached to shipping containers with information regarding the identity of the cargo and potential hazards.

CHRIS, Chemical Hazards Response Information System (U.S. Coast Guard)

Council on Environmental Quality

DERS, Distribution Emergency Response System (Dow Chemical)

U.S. Departments of Commerce, Defense, Interior, State, and Justice

U.S. Department of Transportation
- Office of Hazardous Materials
- The Federal Highway Administration
- The Federal Railway Administration
- Office of Pipeline Safety

EPA's Interagency Radiological Plan

Federal Department of Agriculture, Poison Control Center (Canada)

HELP, Hazardous Emergency Leaks Procedure (Union Carbide)

NATES, National Analyses of Trends in Emergencies System. This system was designed by the Environmental Emergencies Branch of Environment Canada and is a data base on the history of Canadian spills.

NEELS, National Emergency Equipment Locater System (Canadian)

OHM-TADS, Oil and Hazardous Materials Technical Assistance Data System (Canada)

PACE, Petroleum Association for the Conservation of the Canadian Environment. This system is composed of eleven major oil companies.

The Pesticide System Team Network of the National Agricultural Chemicals Association.

TESAC, Transportations Emergency Response System (Allied Chemical)

TERP, Transportation Emergency Reporting Procedure (DuPont)

THI, Transportation Hazards Information (MCA program developed in the late 1960s similar to CHEMCARD but directed more toward hazards)

TWERP, Transportation and Warehouse Emergency Reporting Procedure (American Cyanamid)

Appendix C

FEDERAL STATUTES OF CANADA

1. Canada Shipping Act

Amendments to the Canada Shipping Act (Bill C-2, tabled in the House of Commons on October 19, 1970) provided for the enactment of new regulations covering the pollution of Canadian waters. The maximum fine for the negligent or willful discharge of designated pollutants into Canadian waters was increased from $5000 to $100,000. Legislation was expanded for the prosecution of the ship or owner rather than only for the master of the ship. The new regulations govern such matters as the construction of ships vs. the cargo, their maintenance, manning, qualifications of the ship's personnal, requirements as to pilotage, the manner of ships' operations, type of navigation equipment, provision for pollution control, and cleanup equipment. The regulations also cover such areas as marine traffic control governing such ships, compulsory routing in Canadian waters, and prescribing of the amount of pollutants ships can carry.

The act includes liability provisions for pollution damage. Liability rests on owners of the pollutants. Included in liability is the cost of wreck removal, carbo removal, pollution cleanup, and property damage. Property damage includes compensation for the loss of income for fishermen.

Ship owners are required to put up a bond covering the carriage of pollutant by sea. If an incident occurs without the actual fault or knowledge of ship owners, up to $134 for each ton of ship's tonnage or up to a liability maximum of $14 million has been specified. Where the incident occurs with the actual fault or knowledge of such persons, their liability is unlimited.

A Maritime Pollution Claims Fund exists for unsatisfied judgments with respect to pollution damage. The unsatisfied judgments may arise where the judgment sum actually exceeds the amount to which their liability is limited in cases where the damage occurred without actual fault or knowledge.

Owners of ships carrying hazardous materials and sometimes the owners of the pollutants as well are required to provide evidence of financial responsibility in the form of an indemnity bond or insurance. Failure to do so when required is an offense carrying a maximum fine of $100,000. Failing to report pollution incidents can also result in a maximum fine of $100,000.

2. Fisheries Act

As amended on June 26, 1970,
Rep. and Sub. 1969-70, C. 63, S3

33(2) Subject to subsection (4), no person shall deposit or permit the deposit of a deleterious substance of any type in water frequented by fish or in any place under any conditions where such deleterious substance that results from the deposit of such deleterious substance may enter any such water.

(5) Any person who violates any provision of this section is guilty of an offense and liable on summary conviction to a fine not exceeding $5000 for each offense.

(6) Where an offense under subsection (5) is committed on more than one day or is continued for more than one day, it shall be deemed to be a separate offense for each day on which the offense is committed or continued.

(7) Where a person is convicted of an offense on this section, the court may, in addition to any punishment it may impose, order that person to refrain from committing any further such offense or to cease to carry on any activity specified in the order the carrying on of which, in the opinion of the court, will or is likely to result in the committing of any further such offense.

(11) For the purpose of this section...
(a) "Deleterious substance" means
(1) Any substance that, if added to any water, would degrade or alter or form part of a process of degradation or alteration of the quality of that water so that it is rendered deleterious to fish or to the use by man of fish that frequent that water....

3. The Criminal Code

Section 387

A person commits mischief who willfully:

(a) Destroys or damages property,

(b) Renders property dangerous, useless, inoperative, or ineffective,

(c) Obstructs, interrupts, or interferes with the lawful use, enjoyment or operation of property, or

(d) Obstructs, interrupts or interferes with any person in the lawful use, enjoyment or operation of property.

The punishment for mischief in relation to private property is imprisonment up to five years. If this mischief causes danger to life, the penalty can be life imprisonment.

4. The National Harbours Board Act

4 - (2) Nothing shall be thrown, drained or discharged into the water, allowed to come in contact with the water, or deposited anywhere within the limits of the harbour, which may in any manner
(a) Damage vessels or property; or
(b) Cause any nuisance or endanger life or health;

(3) Every encumbrance, obstruction, nuisance, or possible cause of danger or damage, in contravention of the provisions of this section, may be removed by the board (National Harbours Board) at the risk and expense of the person who so contravenes such provisions.

5. Migratory Bird Regulations

51. No person shall knowingly place, cause to be placed or in any manner permit the flow or entrance of oil, oil wastes or substances harmful to migratory waterfowl on waters flowing into such waters or the ice covering either of such waters.

Penalty. Fines $10 to $300 and/or imprisonment up to six months.

6. Extracts from the Railway Act, Nov. 30, 1962

354- (2)
Every person who sends by the railway any such goods shall distinctly mark their nature on the outside of the package containing the same and otherwise give notice in writing to the station agent or employee of the company whose duty it is to receive such goods and to whom the same are delivered.

355- (1)
The company shall not carry any goods of an explosive or dangerous nature except in conformity with the regulations made by the board on that behalf.

(2)
The company may refuse to take, except in conformity with any order or regulation made by the board in that behalf, any package or parcel that it suspects to contain goods of an explosive or dangerous nature, or may require the same to be opened to ascertain the fact.

440
Every company which carries any goods of an explosive or dangerous nature except in conformity with the regulations or an order, made by the board in that behalf shall for each such offense incur a penalty of $500.

7. The Aeronautics Act

6 - (1) Subject to the approval of the Governor in Council, the Minister (of Transport) may make regulations to control and regulate air navigation over Canada, including the territorial sea of Canada and all waters on the landward side

thereof, and the conditions under which aircraft registered in Canada may be operated over the high seas or any territory not within Canada, and, without restricting the generality of the foregoing, may make regulations with respect to...etc.

(d) the conditions under which aircraft may be used or operated.

(m) the preservation, protection and removal of aircraft involved in accidents, including the cargo thereof... etc.

(o) the investigation of any accident involving an aircraft, of any alleged breach of any regulation made under this section or of any incident involving an aircraft that, in the opinion of the Minister, endangered the safety of persons...etc.

(4) Every person who violates a regulation is guilty of an offence and is liable on summary conviction to a fine not exceeding $5000 or to imprisonment for a term not exceeding one year or both.

(5) Every person who violates an order or direction of the Minister made under a regulation, or who obstructs or hinders an investigation carried on under this act, or the regulations, is guilty of an offence and is liable on summary conviction to a fine not exceeding $1000 or to imprisonment for a term not exceeding six months or to both.

8. International Joint Commission
(United States and Canada)

Under Article IX of the Boundary Waters Treaty of 1909, the IJC has carried out water pollution investigations in the connecting channels of the Great Lakes, the St. Croix River, the Rainy River, the Red River, Lake Erie, Lake Ontario, and the international section of the St. Lawrence River. The term <u>connecting channels</u> includes the St. Mary's River, St. Clair River, Lake St. Clair, the Detroit River, and the Niagara River. The objectives concerning industrial wastes are as follows:

"Adequate protection should be provided for those waters (connecting channels, IJC report 1950).

(g) Substances highly toxic to humans, fish, aquatic or wildlife are eliminated or reduced to safe limits."

Some of the concentration objectives specified are as follows:

	Maximum Permissible
Phenol	5 ppb
Iron	0.3 ppm
pH	range 6.7 to 8.5
DO	not to fall below 3 mg/liter

9. The Ontario Water Resources Act
October 1972 (Provincial Statutes)

30 Under Sections 31, 32, 34 and 36, the quality of water shall be deemed to be impaired if, notwithstanding that the quality of the water is not or may not become impaired, the material deposited or discharged or caused or permitted to be deposited or discharged, or any derivative of such material causes or may cause injury to any person, animal, bird or other living thing as a result of the use or consumption of any plant, fish or other living matter or thing in the water or in the soil in contact with the water. RSO 1970 C332, S. 30

32-(1) Every municipality or person that discharges or deposits or causes or permits the discharge or deposit of any material of any kind into or in any well, lake, river, pond, spring stream, reservoir or other water or watercourse or on any shore or bank thereof or into or in any place that may impair the quality of the water of any well, lake, river, pond, spring, stream, reservoir, or other water or watercourse is guilty of an offence and on summary conviction to a fine of not more than $5000 and on each subsequent conviction to a fine of not more than $10,000 or to imprisonment for a term of not more than one year or to both such fine and imprisonment.

(2) Each day that a municipality or person contravenes subsection 1 constitutes a separate offence. RSO 1970 C 332, S. 32 (1,2)

(3) Every municipality or person that discharges or deposits or causes or permits the discharge or deposit of any material of any kind, and such discharge or deposit is not in the normal course of events, or from whose control material of any kind escapes into or in any well, lake, river, pond, spring stream, reservoir, or other water or watercourse or on any shore or bank thereof or into or in any place that may impair the quality of the water of any well, lake, river, pond, spring, stream, reservoir or

other water or watercourse shall forthwith notify the Ministry (of the Environment) of the discharge, deposit or escape, as the case may be. RSO 1970 C 332, S 32 (3); 1972 C 1, S 70(2)

(4) Every municipality or person that fails to notify the Ministry as provided in subsection 3 is guilty of an offence and on summary conviction is liable to a fine of not more than $5000. RSO 1970, C 332, S 32 (4); 1972, Cl, S 70 (2)

80-(1) In this section "emergency order" means an order, direction, report or notice issued, made or given under this act in an emergency by reason of:

(a) danger to the health or safety of any person;
(b) impairment or immediate risk of impairment of any waters or the use thereof, or
(c) injury or damage or immediate risk of injury or damage to any property or to any plant or animal life.

80-(3) A person or municipality to whom an emergency order is issued, made or given shall comply with the emergency order forthwith after a service of the order or a written copy thereof.

10. The Environmental Protection Act, 1971
(Canada)

3. The Minister (of the Environment) for the purposes of the administration and enforcement of this act and the regulations may:

(a) investigate problems of pollution.
(b) conduct research related to contaminants, pollution, etc.
(c) conduct environmental quality studies.
(d) conduct environmental planning studies.
(e) convene conferences.
(f) gather, publish and disseminate information.
(g) make grants and loans.
(h) establish and operate demonstration and experimental sites.
(i) appoint committees.
(j) contract private agreements related to the conservation of the environment.

5-(1) No person shall deposit in, add to, emit or discharge into the natural environment any contaminant, and no person responsible for a source of contaminant shall permit the addition to, emission or discharge into the natural environment of any contaminant in an amount, concentration or level in excess of that prescribed by the regulations.

15-(1) Every person who
(a) deposits in, adds to, emits or discharges into any part of the natural environment, or
(b) is the person responsible for a source of contaminant that deposits in, adds to, emits or discharges into any part of the natural environment,

out of the normal course of events, any contaminant that

(c) has an offensive odour,
(d) may endanger the health or safety of any person,
(e) may injure or damage or cause injury or damage to
(i) real or personal property, or
(ii) plant or animal life,

shall forthwith notify the Department (of the Environment) of the deposit, addition, emission or discharge, as the case may be.

17. Where any person causes or permits the deposit, addition, emission or discharge into the natural environment of a contaminant that injures or damages land, water, property or plant life, the Minister, where he is of the opinion that it is in the public interest so to do, may order such a person to do all things and take all steps necessary to repair the injury or damage.

18. When in the opinion of the Director (of the Air Management or Waste Management Branch of the Department of the Environment), based upon reasonable and probable grounds, it is necessary or advisable for the protection or conservation of the natural environment to do so, the Director may, by an order directed to any person, require that person to have on hand, and available at all times, such equipment and materials as the order specifies to alleviate the effect of any contamination of the natural environment that may be caused or permitted by the person to whom the order is directed.

41. No person shall use any facilities or equipment for the storage, handling, treatment, collection, transportation, processing or disposal of waste that is not part of a waste management system for which a certificate of approval has been issued and except in accordance with the terms and conditions of such certificate.

42-(1) Where waste has been deposited upon any land or land covered with water or in any building that has not been approved as a waste disposal site, the Director may order the occupant or the person having charge or control of such land or building to remove the waste and to restore the site to a condition satisfactory to the Director.

(2) Where a person to whom an order is directed under subsection 1 fails to comply with the order, the Director may cause the necessary work to be done and charge such person with the cost thereof, which may be recovered with costs in any court of competent jurisdiction.

84-(3) Every person responsible for a source of contaminant shall furnish such information as a provincial officer requires for the purposes of this act or the regulations.

85. Whenever a provincial officer is required or empowered by this act or the regulations to do or direct the doing of anything, such provincial officer may take such steps and employ such assistance as is necessary to accomplish what is required, and may, when obstructed in so doing, call for the assistance of any member of the Ontario Provincial Police Force or the police force in the area where the assistance is required and it is the duty of every member of a police force to render such assistance.

92-(1) Where a person complains that a contaminant is causing or has caused injury or damage to livestock or to crops, trees or other vegetation which may result in economic loss to such person, he may, within fourteen days after the injury or damage becomes apparent, request the Minister to conduct an investigation.

102-(1) Except as otherwise provided in this act, every person, whether as principal or agent, or an employee of either of them, who contravenes any provision of this act or the regulations or fails to comply with an order made under this act is guilty of an offence and on summary conviction is liable on a first conviction to a fine of not more than $5000 and on each subsequent conviction to a fine of not more than $10,000 for every day or part thereof upon which such offence occurs or continues.

11. The Public Health Act (1969)
Province of Alberta

General Sanitation
34-8-2

No person shall discharge into a public sewer any chemicals, chemical substances or their residues, fuel oil or other inflammable substances used in various trades and industries, which might cause damage from explosion or might in any other way prove injurious or dangerous to health.

Approval for Discharge to Surface
Water or Watercourse
39-3-1

(1) No person, corporation or municipality shall, without the written permission of the Provincial Board of Health:

(a) discharge, deposit, drain, release, cause or allow to be deposited upon the banks of or into any reservoir, surface water or watercourse any substance capable of changing the quality of the water or causing water pollution; or

(b) place any pesticide in any reservoir, surface water or watercourse; or

(c) place any pesticide on the bank of any reservoir, surface water, or watercourse unless exemption of permit requirement has been granted by the Provincial Board of Health.

Penalties are relatively low being in the order of $50 to $200.

The following criterion for water has been specified (only highly hazardous substances are listed):

Substance	Criterion, mg/liter
Arsenic	0.05
Cadmium	0.01
Chromium	0.05
Fluoride	1.5
Lead	0.05
Selenium	0.01
Cyanide	0.01
Phenols	0.005
pH	6.5-8.5

12. The Municipal Government Act (1968)

Sanitation and Health

The council (municipal), subject to the provisions of the Public Health Act and any regulations thereunder, may pass by-laws....

(b) preventing or restricting, controlling and regulating the discharge into any stream, watercourse, drain, sewer or sewerage system of any deleterious matter, substance or thing, whether liquid or solid, that would be injurious to health, life or property, or injure, pollute or damage any stream, watercourse, drain, sewerage system or sewage treatment plant.

Penalties are left to the judgment of the courts.

13. The Litter Act (1972, Chapter 61, Alberta, continued)

Part 1, Section 6

1. No person shall dispose of litter on, into or under water or ice....

2. "Dispose" includes spilling, leaking, pouring, emitting and emptying.

Section 7, Offences and Penalties

First offence – $50 to $100
or five days imprisonment
Second offence – $150 to $250
or fifteen days imprisonment
Third offence – A fine of not less than $500 or not more than ninety days imprisonment.

14. Clean Water (General) Regulations
OC 216/73

Part 3, Section 9

1. Where
   (a) Any uncontrolled release of contaminant, or
   (b) Any controlled release of contaminants not authorized by a licence, or
   (c) Any accidental spill or discharge occurs,

   the operator or person in charge of the water facility from which the release, spill or discharge occurs shall report the occurrence to the Director of Pollution Control within twenty-four hours of
   (d) its discovery, or
   (e) notification by another person of the release, spill or discharge, or
   (f) the time the release, spill or discharge should have been discovered if the water facility was being operated or used in an efficient manner.

2. Within 72 hours of a release, spill or discharge, referred to in subsection 1, the owner of the water facility or his agent shall notify the Director of Pollution Control of the occurrence in writing including the following details:
   (a) The date and time of the occurrence,
   (b) The location of the point of the release, spill or discharge,

(c) The composition of the release, spill or discharge showing with respect to each contaminant
  (i) its concentration
  (ii) its emission rate, and
  (iii) the total weight, quality or amount,
(d) A detailed description of the circumstances leading to the release, spill or discharge,
(e) The steps taken to control or stop the release, spill or discharge, and
(f) The steps that will be taken to prevent similar future releases, spills or discharges.

Section 10

Any person who contravenes any provision of Section 9 is guilty of an offence and liable on summary conviction to a fine of not more than $1000 for each day that the contravention continues.

15. Clean Water (Industrial Plants) Regulations OC 217/73

Part 1 — Application for a permit to construct an industrial plant

Subsection 2

An application for a permit must be accompanied by:

....

(j) A description of the potential danger of an emergency and accidental discharge of contaminants and the procedures that will be followed following an emergency or accidental discharge.

16. Clean Air Regulations OC 214/73 (Alberta, continued)

Part 1 — Application for a permit to construct a plant.

Subsection 3

An application for a permit shall contain the following information where applicable, with respect to the plant:

....

(l) A description of the contingency procedure that will be taken to prevent the discharge of untreated wastes in the event of a power failure, mechanical failure of the pollution control facilities or any failure of the plant's manufacturing equipment.

17. Clean Air (General) Regulations
OC 215/73

Part 3, Returns and Reports

Section 9

1. Where
   (a) Any uncontrolled release of an air contaminant, or
   (b) Any controlled release of an air contaminant not authorized by a licence, or
   (c) Any accidental release or discharge of an air contaminant occurs, the operator or person in charge of the plant from which the release or discharge occurs shall report the occurrence to the Director of Pollution Control within twenty-four hours of
   (d) Its discovery, or
   (e) Notification by another person of the release or discharge.

Subsection 2 specifies what details are required in this report.

BRITISH COLUMBIA

18. Pollution Control Act (1967)

Section

10. The director (Director of Pollution Control) has all the powers necessary for carrying out the intent of this act and, without limiting the generality of the foregoing, has power
   (a) To determine what qualities and properties of water, land or air shall constitute a polluting condition;

   ....

   (h) To determine his own procedure;
   (i) To exercise any of the powers or duties conferred or imposed upon an engineer under this act or the regulations.

(1970)

11. In addition to all other powers given under this act and the regulations, every engineer
    (a) May determine what constitutes a substantial alteration or impairment of the usefulness of land, water or air;
    (b) May enter at any time and upon any land and premises to inspect, regulate, close, or lock any works or premises; and
    (c) May order the repair, alteration, improvement, removal of, or addition to any works.

(1970)

20A. Every person who is guilty of an offence against this act is liable, on summary conviction, to a fine not exceeding $1000 or to imprisonment for a term not exceeding three months, or to both, and if the offence is of a continuing nature, to a fine not exceeding $500 for each day the offence is continued....

19. The Health Act (1967)

Regulations

The Lieutenant-Governor in Council may make and issue such general rules, orders and regulations as he deems necessary for the prevention, treatment, mitigation, and suppression of disease, and may from time to time alter or repeal any such rules, orders and regulations, and substitute new rules, orders and regulations, and the Lieutenant-Governor in Council may by the rules, orders, and regulations provide for and regulate....

....

(r) The prevention of the pollution, defilement, discoloration, or fouling of all lakes, streams, pools, springs and waters;

71. Any person who violates an order, direction, by-law, or regulation of a Local Board, made pursuant to the act or these regulations, shall be liable, on summary conviction, under the Summary Convictions Act, for every such offence to a fine not exceeding $100, with or without costs, or to imprisonment, with or without hard labour, for a term not exceeding six months or to both fine and imprisonment, in the discretion of the Convicting Justice.

MANITOBA

20. The Clean Environment Act (1968)

Notwithstanding any other Act of the Legislature without a subsisting licence from the corporation, no person either directly or indirectly,

....

(b) shall leave, deposit, or throw, or permit or cause to be left, deposited, or thrown, any manure, night soil, decayed or decaying matter or other sewage or the carcass or offal of any animal or fish or part thereof, or any lime, chemical substances, drugs, poisonous matter, garbage, refuse, cans, bottles, rubbish, or any other filthy or impure matter of whatsoever kind, within two chains of the normal high water mark on any body of water or into the waters of, or upon the ice of, any body of water in the metropolitan area or the additional zone.

Regulations for Metropolitan Winnipeg

2. Except as hereinafter provided by licence, at an annual fee of $25, no person shall discharge or cause to be discharged any of the following described kinds of sewage, industrial or factory wastes, into any sewer or body of water within or entering the metropolitan area or additional zone:

....

(g) Any water or waste having a pH lower than 5.5 or higher than 9.0 or having any other corrosive property capable of causing damage or hazard to structures, equipment, and personnel of the sewage works;

(h) Any water or waste containing toxic or poisonous substance

(i) Any noxious or malodorous substance capable of creating a public nuisance

....

(k) Any industrial wastes whatsoever.

21. The Public Health Act

Part III — Division 7
Protection of Water Sources

59. No person, firm, company or corporation shall discharge any raw or untreated sewage, creamery, trade or mine waste into

any water course of such a nature that will, or may create any dangerous or offensive condition and nuisance or impair or render any waters used for any municipal or private domestic supply dangerous or unfit for use, nor without the written permission of the Minister.

22. The Municipal Act

Regulations Concerning Public Health and Welfare

(3) Any municipal corporation may pass by-laws not inconsistent with the Public Health Act and the regulations thereunder,

(a) For preventing and abating public nuisances,

....

(d) For regulating or preventing the encumbering, injuring, or fouling by animals, vehicles, vessels, or other means, of any public wharf, dock, slip, drain, sewer, river, or water, or any highway.

Penalty..... A fine not exceeding $50, or he may be imprisoned with or without hard labor, in the first instance for a term not exceeding 30 days....

PROVINCE OF QUEBEC

23. Water Board Act (RSQ 1964, C 183)

Regulation No. 1 - April 1969

Section 1

The deposit or discharge of organic or inorganic matter which because of its very nature or because of the quantity of the deposit or discharge, renders unclean all or a part of a sheet of surface or subterranean water for human, animal or industrial consumption, or which makes such water injurious to the life of fauna and flora or to the practice of sports, constitutes an operation entailing the pollution of water, and is prohibited by this regulation.

Offences and Penalties

Whosoever infringes any provision of this act or of the regulations shall be liable, on summary proceeding, for the first offence, to a fine of $25 to $100 and, for each subsequent offence within twelve months, to a fine of $100 to $500.

24. The Cities and Towns Act (RSQ 1964, C. 193; am. 1966-7, C 54; am. 1968, C. 55; am. 1969, Bill 13)

The council (of a city or town) may make by-laws:

(1) To define what shall constitute a nuisance and to abate the same, and to impose fines upon persons who may create, continue or suffer nuisances to exist. (RSQ 1964, C 193, S492)

NEW BRUNSWICK

25. The Water Act (1969)

Water Pollution — General Prohibitions

Unless approved by the Minister (of the Environment) and the Minister of Health and Welfare, no municipality or person shall discharge or deposit any material of any kind into or in any well, lake, river, pond, spring, stream, reservoir, or other water or watercourse or upon any ice lying therein or thereon or on any shore or bank thereof or into or in any place that may cause pollution or impair the quality of the water for beneficial use.

....

(2) In such defined and prescribed area no person shall

(a) place, deposit, discharge or allow to remain therein any material of any kind that may impair the quality of the water;

NOVA SCOTIA

26. Public Health Act (1967)

(1) No person shall ....

....

(b) cause, suffer or permit any sewage, refuse, garbage, rubbish or any matter to discharge or flow on or adjacent to any street or public road or into any drain, sluice, or watercourse on any street or public road.

(2) A person who violates this section is liable in addition to a penalty to payment of the expense of removing such

sewage, refuse or other matter or of preventing such discharge or flow. (R.S.N.S. 1967, C.247, S.47)

Water Pollution as a "Nuisance"

(1) No person shall put in any place, on land or water, any offensive matter or thing likely to endanger the public health or likely to become a nuisance.

Offences and Penalties

Any person who violates any provision of this act or regulations made under it shall, unless a penalty is otherwise specially provided, be liable on summary conviction to a penalty not exceeding $100. (R.S.N.S. 1967, C.247, S.128)

27. Environmental Pollution Control Act (1970)

4. (1) Where by reason of any act or omission over which the Legislature of Nova Scotia has jurisdiction, pollution has occurred or is likely to occur that is harmful to the public health, safety or welfare, the Ministry may direct the Commission to investigate and make a report to the Minister.

(2) Where the report made to the Minister recommends that it is in the public interest to require the taking of actions to remedy the act or omission that has caused or is likely to cause the pollution and the Minister approves the report, the Commission may make an order requiring any person or persons to take remedial action to combat, eliminate or mitigate the cause of the pollution.

Appendix D

A DEPLOYMENT MODEL FOR THE RETENTION OF A SPILL OF A HIGHLY HAZARDOUS MATERIAL AND CARBON ADSORPTION DATA

D-1 DEPLOYMENT MODEL FOR AN ACCIDENT INVOLVING A LEAKING CARRIER THAT IS SUBMERGED IN WATER

A tank truck carrying hazardous material has been submerged in water. The vehicle has been located and marked and is leaking hazardous substance into the waterway. The action steps are as follows:

1. The barrier arrives at the spill scene by pickup truck.
2. Three men unload the barrier into the water from the truck, while two other men are unloading a small boat equipped with a davit from a second truck that has arrived with auxiliary equipment.
3. Thirty minutes have elapsed since the deployment team arrived at the spill scene. The small boat with two men aboard is now towing the barrier from shore to the leaking vehicle. The other men are unloading the remaining equipment and preparing explosive anchors for deployment.
4. The men in the small boat are now positioning the barrier above the spill source. Sections have been lifted over the marker buoys and temporary anchors are being attached to the anchor points on the barrier.

5. The barrier is now positioned in a taut pentagonal shape above the spill source by adjusting the five temporary anchors.
6. One hour has elapsed and the small boat returns to shore and picks up the first explosive anchor with its davit.
7. The first explosive anchor is transported to the moored barrier and is lowered to the bottom at a point directly under an anchor point on the barrier. The anchor point has a marker buoy attached so that it is easily located.
8. The first anchor is fired into the bottom. The firing gun is recovered and the wire rope pendant from the explosive anchor is temporarily attached to the anchor point on the barrier. The marker buoy supports the weight of the wire rope and prevents the barrier from sinking. The remaining four explosive anchors are installed in similar fashion.
9. Two and one-half hours have elapsed since the arrival of the deployment team at the spill site. The boat has now picked up five pull-down devices from shore and is returning to the barrier.
10. Attachment of the first pull-down device begins. First the wire rope pendant is detached from the anchor on the barrier and the eye on the end is cut off with a wire cutter. Next the pull-down grip and mooring grip are slipped on the cable. The cable is then put in a cable grip attached to the davit and pulled taut, after which the pull-down grip is allowed to fall down the wire rope to the bottom. The pendant is then slackened and the mooring grip is attached to the anchor point. The remaining four devices are installed using the same procedure.
11. Three and one-half hours have now elapsed. Upon installation of the fifth pull-down device the anchor pendant is left in the cable grip and the barrier is pulled to the bottom. The operation is repeated at each anchor point. After each anchor point has been pulled to the bottom, the

ends of the wire rope pendant and the pull-down rope are attached to the marker buoy that has been removed from the anchor point.

12. The small boat returns to shore and picks up the air compressor-water pump. A total of four and one-half hours have elapsed as the inflation of the barrier begins. As air is pumped down to the barrier it gains buoyancy, breaks the tape bindings and begins to rise to the surface. The rip cord is pulled to break the remaining bindings in order to shorten the time required for the barrier to rise to the surface.
13. Five hours after deployment began, the barrier has risen to the surface and is effective in containing the leaking material. Inflation continues for another forty-five minutes until the flotation collar is completely inflated.
14. Inflation of the water bladder begins and at an elapsed time of seven hours after initiation of deployment the barrier is in place with all bladders inflated, and deployment is complete.

The above operation took place under ideal conditions. Under adverse conditions more than 7 hr would be required.

Source: L. S. Brown, A Physical Barrier System for Control of Hazardous Material Spills in Waterways, Proc. 1972 National Conf. Control of Hazardous Material Spills, Houston, Texas, March 21-23, 1972.

D-2 ACTIVATED CARBON TREATMENT FOR CONTAMINATED WATER

Purpose of test work: To test the feasibility of treating contaminated water directly with activated carbon which can be later removed and incinerated.

Test procedure: 100-ml samples of contaminated water were agitated with added amounts of activated carbon for a given period of time. The mixture was filtered and the filtrate tested for remaining contaminant. Spectral photometry, gas chromatography, and atomic absorption spectroscopy techniques were used.

Experimental results: Phenol adsorption is shown in Figure D-2-1. Removal efficiency increases with higher initial concentrations of phenol.

Heavy metal adsorption is shown in Table D-2-1. Mercuric ions are very effectively adsorbed by activated carbon. Lead removals were also promising. Copper and chromium removals were not so effectively adsorbed.

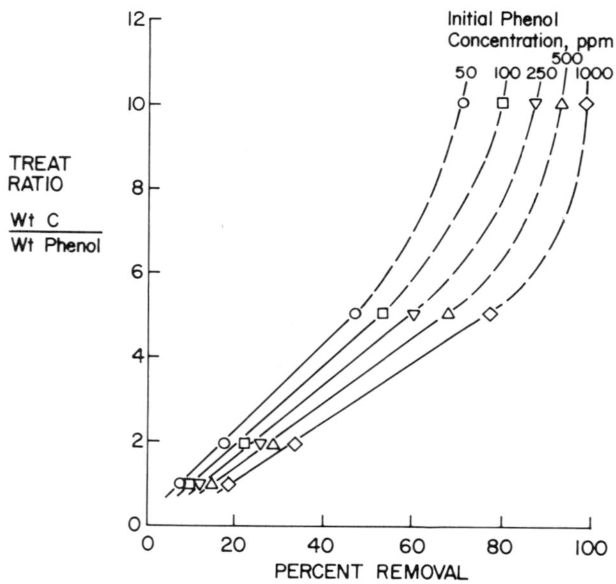

Figure D-2-1 Removal of phenol from contaminated water by carbon adsorption. Source: R. C. Ziegler and J. P. La Fornara, In Situ Treatment Methods for Hazardous Material Spills, Proc. 1972 National Conf. on Control of Hazardous Spills, Houston, Texas, March 21-23, 1972.

Table D-2-1

Removal of Heavy Metal Ions from Contaminated Water[a] by Carbon Adsorption[b]

Heavy Metal Tested	Activated Carbon Dosage, ppm	Initial Metal Concentration, ppm	Residual Metal Concentration, ppm	Removal, %
Hg as $HgCl_2$	0	100	100	0
	500	100	1	99
	1000	100	1	99
	10000	100	1	99
Cu as $CuSO_4$	0	50	50	0
	500	50	46	8
	1000	50	45	10
	5000	50	13.5	73
	10000	50	1.8	96.4
Cr as $CrO_3 \cdot 6H_2O$	0	100	100	0
	500	100	95	5
	1000	100	92.5	7.5
	5000	100	82.5	17.5
	10000	100	52.5	47.5
Pb as $Pb(NO_3)_2$	0	107	107	0
	500	107	95	5
	1000	107	88	17.7
	5000	107	17	84
	10000	107	7.5	93

Source: R. C. Ziegler and J. P. La Fornara, In Situ Treatment Methods for Hazardous Material Spills, Proc. 1972 National Conf. on Control of Hazardous Material Spills, Houston, Texas, March 21-23, 1972.

[a]No pH adjustments.

[b]Carbon: Aqua Nuchar C-190N, West Virginia Pulp and Paper Co.

Appendix E

DEFINITION OF NUCLEAR TERMS

CURIE

One curie (Ci) is that amount of radioactive material in which 3.4×10^{10} atoms disintegrate per second. The activity of 1 g Ra-226 together with its decay products equals one curie. The specific activity of a material is expressed in curies per gram or mCi per mg. Pure radium has a specific activity of one Ci/g; that of pure P-32 is about 50 μCi/g.

About 1.8×10^6 Ci of radioactive fission products are contained in a reactor core per MW of power one hour after shutdown after a two-year operating cycle (see Figure E-1).

Activity in Ci does not express the dosage of radiation, only the number of ionizing particles emitted independent of their energy or range.

DOSAGE
(rem, roentgen equivalent man)

One rem is the quantity of radiation that produces the same biological effects in man as those resulting from the absorption of one roentgen of x-ray or γ-radiation.

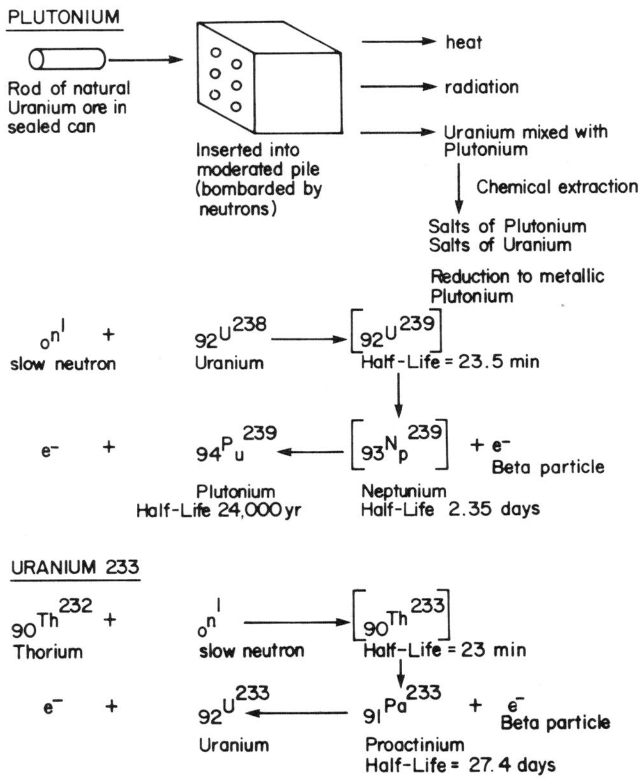

Figure E-1 Fissionable byproducts.

One roentgen is that quantity of x-ray or γ-radiation that produces one esu/ml of dry air at stp (standard temperature and pressure). One esu is equivalent to 1.61×10^{12} ion pairs per gram of dry air at stp. One ion pair requires about 32.5 eV. The roentgen was adopted by the radiological congress of 1937.

During one roentgen of radiation, 83×10^{-7} of energy is absorbed per gram of dry air at stp. In terms of eV, this would be 5.23×10^{13} eV per gram of air. Soft body tissue is a better absorber of radiation and one roentgen produced an equivalent of 93×10^7 J per gram of tissue.

At present, the maximum permissible dosage (mpd) is 5 rem per year for industrial workers. Individual members of the general population are allowed 0.5 rem/year, and masses of population 0.17 rem/year.

HALF-LIFE

The time period taken for half the radioactive atoms in a given mass of an element to disintegrate. In one gram of Ra-226 approximately 3.61×10^{10} disintegrations occur every second and the half-life is 1590 years. One gram of U-238 releases 12,000 α-particles per sec and has a half-life of 4.6×10^9 years.

IONIZATION

Specific ionization is the number of ion pairs produced per unit length of a particle's track through a given medium. Common reference media are air and water and expressions are in terms of ions/ml for air and ions/μml for water.

Total ionization is the number of ion pairs produced in a medium as the incident particle comes to a halt in this medium. In general, for dry air at stp:

$$\frac{\text{particle energy in eV}}{32.5} = \text{number of ion pairs}$$

FAST NEUTRON REACTOR

In a fast neutron reactor most of the fissioning is caused by fast neutrons. The physical reactor size is smaller for a given output. Fast neutron reactors have more possibilities as breeders which in part are necessary for fissionable weapon cores. The cooling of <u>fast</u> reactors is more complex than of <u>slow</u> reactors.

SYMBOLS

MW(e) = megawatt electrical

m = 10^{-3}

μ = 10^{-6}

INDEX